THE
ORDNANCE SURVEY
PUZZLE BOOK

THE
ORDNANCE SURVEY
PUZZLE BOOK

Pit your wits against Britain's
greatest map makers

**By Ordnance Survey
and Dr Gareth Moore**

Mapping images © Crown Copyright and database rights 2018
Text and other images © Ordnance Survey Limited 2018
Puzzles © The Orion Publishing Group Ltd 2018
Puzzles by Dr Gareth Moore

The right of the Ordnance Survey to be identified as
the authors of this work has been asserted in accordance with the
Copyright, Designs and Patents Act 1988.

This edition first published in Great Britain in 2018 by
Trapeze
an imprint of the Orion Publishing Group Ltd
Carmelite House
50 Victoria Embankment
London EC4Y 0DZ
An Hachette UK Company

5 7 9 10 8 6

A CIP catalogue record for this book is available
from the British Library.

ISBN: 978 1 4091 8467 6

Printed in Italy

The names OS and Ordnance Survey and the OS logos are protected by UK trade
mark registrations and/or are trademarks of Ordnance Survey Limited, Great Britain's
national mapping agency.

Every effort has been made to fulfil requirements with regard to
reproducing copyright material. The author and publisher will be
glad to rectify any omissions at the earliest opportunity.

www.orionbooks.co.uk

MIX
Paper from
responsible sources
FSC® C023419

CONTENTS

PUZZLES

FOREWORD

For more than 225 years, Ordnance Survey (OS) has been mapping Great Britain and we've seen the landscape change dramatically over this period as the nation has evolved. Considering that the population of Great Britain in 1791 (the official year of OS's incorporation) was estimated as a little over 8 million, there has been significant development in our towns and cities to support a growth in population of just under 60 million. Our maps soon became an iconic reminder of our history, and the amount of changes that take place across Britain's landscape every day.

Throughout our existence, our maps have inspired fondness, fuelled ambition and led to a certain amount of obsession. To some they are for scribbling on, to track where they've been or to highlight favourite destinations; to others they are collector's items, spending most of their time gathering dust on a shelf. As a nation, we use maps to get from A to B, to plan and to navigate, but without having the time to appreciate some of the smaller intricacies tucked away in other parts of the map.

We are still best known for our pink (Landranger) and orange (Explorer) leisure maps; but today Ordnance Survey is predominantly a digital company. The majority of our work is in helping cities, governments and companies run efficiently, as everything happens somewhere, and we know that 'somewhere' – which in some cases can be critical. But the leisure element of our business is still as vital today as it ever was, in fact possibly more so. Our digital mapping is included in pretty much every map of Great Britain and also in the OS Maps app. We hope that by embracing technology through 3D visualisation and augmented reality and showing people the routes they can take, we are helping more people to get outside and explore the countryside.

We see our core purpose within our leisure business as helping more people to get outside more often. Every map can inspire adventure and I'm sure that this book will too. While spending time solving these puzzles will certainly exercise your brain, don't forget to keep the body active too. We hope we have given you some fuel to inspire you. Great Britain is a beautiful and varied country with a variety of landscapes and wildlife just a short distance away from where you live and work.

We hope that this book will give you some insight into these peculiarities or at the very least inspire you to take a closer look the next time you pull a map from your book shelf.

Get inspired, get exploring and, above all, get outside!

MANAGING DIRECTOR – ORDNANCE SURVEY LEISURE.

TIMELINE OF ORDNANCE SURVEY

1747 William Roy, an innovative young engineer, creates a military map of Scotland. But he has a bigger vision: to survey and map the whole of Britain.

1791 With Europe in turmoil during the French Revolution, the British government orders its defence ministry – the Board of Ordnance – to begin surveying England's vulnerable southern coasts. The board buys a huge new Ramsden theodolite for the most precise mapping possible. Ordnance Survey is born.

1791 The triangulation of Britain begins, using a baseline at Hounslow Heath mapped out by William Roy seven years earlier.

1801 The first 1 inch maps are published for the 'invasion coast', starting with Kent.

1810 The Ordnance Survey name appears for the first time – on the 1-inch map of the Isle of Wight and Port of Hampshire.

1841 Height is introduced to Ordnance Survey maps, with levels in relation to a datum point at Liverpool. The Survey Act grants surveyors the right to enter private property.

1855 Photography is introduced to map printing, with zinc plates to replace lithography (stone); 1 inch maps are engraved on copper plate until the 1920s.

1902 Women are employed for the first time at Ordnance Survey – to mount and colour maps.

1914 During the First World War, theodolites are used to work out the positions of enemy guns and aircraft, and to map trenches. A temporary factory set up in France provides the Allies with 3 million maps.

1918 The twentieth century brings more cyclists and motorists on to the roads and ramblers into the countryside.

Professional artist Ellis Martin is appointed to produce eye-catching covers for the 1-inch maps. His classic designs boost sales to record levels, and maps are soon seen as essential by the general public.

1935 The re-triangulation of Great Britain begins – 6,500 trig pillars are built on inhospitable peaks as solid triangulation points. Many of these are still in the landscape today. The principle of 'continuous revision' to keep maps up-to-date is introduced.

1939–45 During the Second World War, Ordnance Survey is the principal supplier of maps for Europe and Africa, including mobile printing operations.

1940 Ordnance Survey HQ is destroyed in the Southampton Blitz, but map supplies are unaffected. Perhaps the most precious item lost is the Ramsden theodolite used in the first surveying work.

1974 The Pathfinder (1:25 000 scale) is used to create Outdoor Leisure Maps. The 1-inch map is replaced by the metric 1:50 000 map called Landranger.

1974 Light-beam technology and automatic data recording equipment allow us to collect data more rapidly.

1978 More than 100 staff are employed just for digitising maps..

2001 OS MasterMap® is launched – a consistent and maintained database that records more than 500 million man-made and natural features of Great Britain in one continuous digital map.

2003 OS Net is introduced to provide a national coordinate reference system which can achieve accuracy levels of 1–2 centimetres.

2010 OS OpenData is launched, giving access to free, unrestricted OS mapping.

2015 OS Maps – the award-winning app – is created.

2017 OS data plays a role in new technologies such as driverless cars and Smart Cities.

The Future Location data relevance increases dramatically to support Fourth Industrial Revolution technologies.

WHO WE ARE AND WHAT WE DO

Who we are

Ordnance Survey (OS) is Britain's mapping agency. While our much-loved paper maps are still popular, we're now a world-leading digital data organisation.

No one knows Great Britain quite like we do. We map every square kilometre of it – surveying every fixed physical object from the ground upwards. Because our data is so precise and accurate, it's used in many ways across all sectors.

Creating detail

Our data capture techniques help us to add 20,000 features to our database each day. These include a team of around 200 surveyors, and aerial photography from an extensive flying programme. Our field surveyors now capture information using OS Net, our national network of Global Navigation Satellite System base stations – the modern equivalent of concrete trig pillars.

We're exploring new ways to capture even more detailed features in the landscape, including vehicle-based surveying on the ground and using unmanned aerial vehicles (also known as drones).

Working with business

Right now, many businesses are using our data to improve the services they provide their customers. It helps locate and manage underground assets and improve service coverage, for example. Insurance companies use it to calculate more accurate premiums and telecommunications companies need an accurate understanding of the landscape for effective mobile networks.

We help to keep transport flowing and ensure online shopping gets delivered to the door, and our data underpins every property sale in Great Britain.

Supporting government

We're a government-owned organisation, providing a hugely valuable service whereby every arm of government can use our resources free of charge. This not only saves tens of millions of pounds, but lives too. Emergency vehicles have the most up-to-date mapping on board, which means they can find the fastest way to an incident using our routing information, and look up addresses and access points with pinpoint accuracy.

We also help organisers plan and manage major events such as the Olympics, Paralympics and Commonwealth Games – keeping everyone safe and secure.

When disaster strikes, such as severe flooding or threats to national security, our accurate mapping is crucial. Everyone using the same source of information means that first responders and central and local governments can distribute their services effectively. Knowing about hazardous terrain, current roadworks and the height of a building can be the difference between life and death in a crisis.

The government also uses our data to improve services in the community – to mend that broken street light quickly, collect our rubbish on time and improve transport networks.

Now we're taking our expertise worldwide. We're working with different countries, showing how location data can help build better economies and communities.

Working with partners

We have around 400 business partners who are licensed to use, enhance and distribute our data, so millions of customers can access and get more value from it. Our partners range from corporate multinationals to small innovative startups across sectors including land and property; finance and insurance; energy and infrastructure; government; retail; transport and logistics; computer gaming; advertising and analytics. They can supply our data, provide the infrastructure or software to make it work or, in many cases, both.

Keeping Britain active

For more than 200 years, Ordnance Survey has helped people explore Great Britain. Today, we're inspiring even more people to get outside more often. We build content and tools to make it easier for everyone to have an active outdoor lifestyle – helping people live longer, stay younger and enjoy life more.

Our iconic leisure maps are available on paper, mobile and desktop to make the outdoors enjoyable, accessible and safe. Every day we're developing innovative new navigation features, including groundbreaking 3D mapping, augmented reality guides and GPS devices. With the rapidly growing GetOutside initiative, endorsed by the Outdoor Industry Association, we're inspiring people to find new places, create lasting memories and discover the great British countryside.

OS Maps is our award-winning app for exploring the great outdoors. Combining Ordnance Survey's iconic mapping (1:25 000, 1:50 000) with other features such as 3D and augmented reality, the app can be used across devices to view, create and sync routes between desktop and mobile.

OS Maps subscribers can access all 607 of our paper leisure maps online. Favourite areas can be printed or downloaded, so any part of the country can be viewed even without a mobile signal.

As well as 1.8 million existing routes (which include several thousand premium routes from walking and hiking magazines), users can plot their own to follow, print and share.

More recently we've introduced an aerial 3D layer, which lets people 'fly through' their routes in stunning 3D before setting out.

Users can also view OS Maps in augmented reality mode – which gives additional leisure information about the local area.

OS IN NUMBERS

We add **20,000** features to our database every day

Our database holds **500** million geospatial features on England, Scotland and Wales

Our surveying equipment is capable of achieving up to **1–2cm** accuracy, giving the most precise picture of the nation

Our aircraft capture more than **150,000** aerial images every year

We receive **62,000** open data orders every year

We collect data for all **243,241sq km** of Great Britain

Our data is used by more than **4,700** organisations working for the public good

We've mapped more than **1,800,000** routes in our OS Maps app

90+ orbiting GNSS satellites provide OS with geospatial information

1791: The year OS was founded

We've mapped **584,008km** of rivers, waterways and canals in our comprehensive water network dataset, helping government agencies and environmental organisations in their management, stewardship and conservation

We fulfil on average **140,000** customer data orders a year

Today's geospatial industry supports some **£26bn** of the UK economy and OS revenue represents c. **12** per cent of the **£1.1bn** UK market for geospatial data

Great Britain's three longest coastlines are: Cornwall **1086km**; Essex **905km**; and Devon **819km**

We produce **607** paper leisure maps, each covering a different area of the whole of Great Britain

Snowdonia is the area with the most routes in our OS Maps app – with **3,779** routes running through a single 1km grid square

AN INTRODUCTION TO THE MAPS AND PUZZLES

The maps in this book have been selected for a significant reason – whether that is a noteworthy moment in national history, a rare geographical feature or a site of special cultural interest – and you will find descriptions, along with the map source, accompanying each one to explain.

Then there are the questions. The puzzles are designed to make you look a little closer and dig a little deeper into the detail of each map, unearthing what you might have missed at first glance and testing how far your skills will take you.

There are a mix of word puzzles, search-and-find clues, general knowledge questions, connection-making conundrums and various mathematical, map skill and navigational challenges throughout. These are split into four levels of difficulty:

- **Easy**
- **Medium**
- **Tricky**
- **Challenging**

You can use the working out pages to jot down your methods and answers, and you'll find solutions from page 195. There is something for everyone to have a go at, but the question is – will you amble your way through or will you push the boundaries to become the ultimate mapmaster? Set off on your puzzle journey to find out!

PUZZLES

OS Fact: Minecraft

In 2013, we used OS OpenData to recreate the map of Great Britain in *Minecraft*. It's the only OS map to win a place in the *Guinness World Records* (as the largest real-world place represented in *Minecraft*). Originally created with 22 billion blocks representing the 243,241sq km of the country, the map was updated in 2014 with more detail, taking it to a staggering 83 billion blocks. We used 1:25 000 scale OS VectorMap District for a smoother, more expansive appearance that's closer to real life.

DEFENCE OF THE REALM

MAP
I

AN ENTIRELY NEW AND ACCURATE SURVEY

The map: 1801 General Survey of Kent

The story: Every letter and line engraved by hand

It was the military's need to 'see what was over the next hill' and to plan defensive positions or outflank the enemy that ultimately gave us the modern map. This 1801 map of Kent, produced by the Board of Ordnance, became the first map of Ordnance Survey. This 'General Survey of England and Wales – An entirely new & accurate Survey of the County of Kent' – as inscribed into the corner of the map – was produced first because of fears of a French invasion. Our small extract of the map shows the North Downs above Seven Oaks (now Sevenoaks). Hachures (the hairy hills) were used to indicate slopes. Spot heights and contours only came into use on OS maps of Ireland in the 1830s and in Britain after the determination of mean sea level at Liverpool in 1844.

▨ Easy

1. How many mills are labelled on the map?

2. Which location on the map is one letter away from the name of the oldest university in the English-speaking world?

▨ Medium

3. Which location on the map shares its name with a biblical vessel?

4. What do 'Hall', 'Briton' and 'Preston' have in common?

▨ Tricky

5. The name of which US state capital appears on the map?

6. Which unit of time appears on the map, and which location contains the number of sub-units in one of those units of time?

▨ Challenging

7. Which locations on the map could potentially mean the following?
   a. A rabbit's place of residence
   b. Greyish-brown booth
   c. Directs elevation
   d. Sun garden

8. How many two- or three-word place names on the map begin with an adjective? For example, 'great' in 'Great Town', were it a place on the map?

MAP
2

WHERE TO DRAW THE LINE

The map: 1 inch (1965) enlarged to 1:50 000 scale with overprint

The story: General Roy's Survey Base

The 1783 triangulation between London and the Channel coast had hoped to answer concerns between the English and French astronomical observatories over their locations on the surface of the Earth. The results were inconclusive but proved the value of triangulation as a technique for measuring accurate distances and angles. The 1784 remeasurement of William Roy's baseline across Hounslow Heath employed 30 metre metal chains, which had proved to be as good as the glass rods Roy had used for the 1783 measurement and more convenient. Accurate to 1 inch in 27,404 ft, this line was the start of the first national triangulation of the country and the foundation of every Ordnance Survey map since. Today, the ends of the line are marked with a half-buried upright barrel of a cannon, with one now near the entrance to Heathrow Airport. The line was shown on early 1 inch mapping but has been added to this later edition with the airport as a blue line.

◻ Easy

1. What is the highest number that appears on the map?

2. What is the greatest number of stations that is marked on a single train line?

◼ Medium

3. Pair up the following shorter words to make longer words that are found on the map: bed, bury, chat, font, hat, kempt, on, sun, tern and ton.

4. What's the greatest number of schools marked in a single row of the grid?

◼ Tricky

5. Which words on the map could potentially mean the following?
   a. Argument on a moor
   b. Silvery-grey car
   c. Value of a certain *Star Wars* protagonist
   d. Touched meat

6. With reference to the map, which of these words is the odd one out: 'ford', 'slow', 'ton' and 'worth'?

◼ Challenging

7. Given that the scale of the map is 1:50 000, how far is the historical temple site from Sunbury Court as the crow flies, to the nearest kilometre?

8. Can you find places on the map whose names are anagrams of the following phrases? Ignore the spaces and punctuation, which may differ from those in the place names.
   a. DRY, HOT HEN
   b. BLIND GIN QUEUES
   c. FROM MACHO DONS

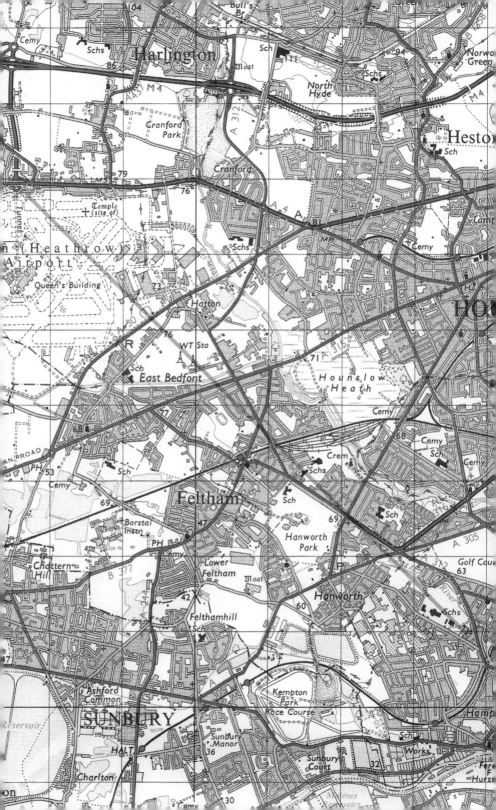

MAP
3

COUNTERING THE ZEPPELIN THREAT

The map: Landranger enlarged to 1:40 000 scale

The story: Best-preserved WWI Airfield in Europe (built 1916)

Lighter-than-air machines, or airships, had become popular at the beginning of the twentieth century, with those developed by Count von Zeppelin providing large luxury gondolas slung beneath impressive 190-metre rugby ball-shaped balloons. As the First World War started the German military identified the reconnaissance potential that airships offered as they were able to operate over the front line in relative safety. Once it became clear that the war wasn't going to end quickly, they were used to bomb British cities, flying over the North Sea from the northwest German coast. The early raids were very successful and caused a lot of damage as well as civilian casualties. War spurred the rapid development of heavier-than-air machines operating from grass airfields like the one at Stow Maries in Essex. While this successfully removed the 'Zeppelin' airships from the war, it also led to the use of aircraft as bombers and, by the end of the WWI, over 1,400 British citizens had been killed in air raids.

QUESTIONS

◻ **Easy**

1. Can you find a name on the map that is spelled like the plural of a country?

2. How many telephone boxes are shown on the map?

◼ **Medium**

3. Can you find a farm whose name is only two letter changes different to 'Hollywood'?

4. If you were driving from Rookery Farm to Corporation Farm via the most direct route, how many right turns on roads would you make?

◼ **Tricky**

5. Where on the map sounds like it is a pit for an alcoholic spirit?

6. As the crow flies, where is the nearest place of worship to White Elm Farm shown on the map?

◼ **Challenging**

7. How many times does the word 'great' appear on the map?

8. Given that the scale of the map is 1:40 000, what is the distance as the crow flies between the Aerodrome Museum and Marsh Farm Country Park, to the nearest kilometre?

MAP
4

GHQ STOP LINE

The map: Explorer enlarged to 1:15 000 scale

The story: WWII pillboxes

At the start of the Second World War the Dunkirk evacuation had heightened the expectation of a German invasion. With the British Army having to abandon much of its mobile equipment in France, a fixed system of defensive lines was drawn up to compartmentalise the country and delay any invading forces long enough for mobile forces to counterattack. Over 50 defensive lines were created around Britain. After the coastal defences, the General Headquarters (GHQ) Line was the most important, designed to protect London and the industrial heart of the country, ringing cities and using features such as rivers, railways and canals. Important bridges and crossing points were defended by tank traps and concrete pillboxes (not shown on map). Many examples remain across the country, with a typical example still in place at Freewarren Bridge over the Kennet and Avon Canal.

Easy

1. How many locks are marked on the map?

2. Can you find places whose names fit the following patterns, where an underline represents a single missing letter? Spaces have been removed, to make it trickier.
   a. _R_F_O_L_C_S
   b. D_D_D_W_
   c. _I_T_N_A_E_

Medium

3. What is the height of the highest point marked on the map?

Tricky

4. What do the words 'croft', 'graft' and 'wilt' have in common?

5. How many different types of public access route – routes marked in green – are marked on the map?

Challenging

6. Given that the scale of the map is 1:15 000, what place is approximately 1,425 metres west and 1,275 metres south of the Wilton phone box?

7. Which phrase on the map could be synonymous with 'grin machines'?

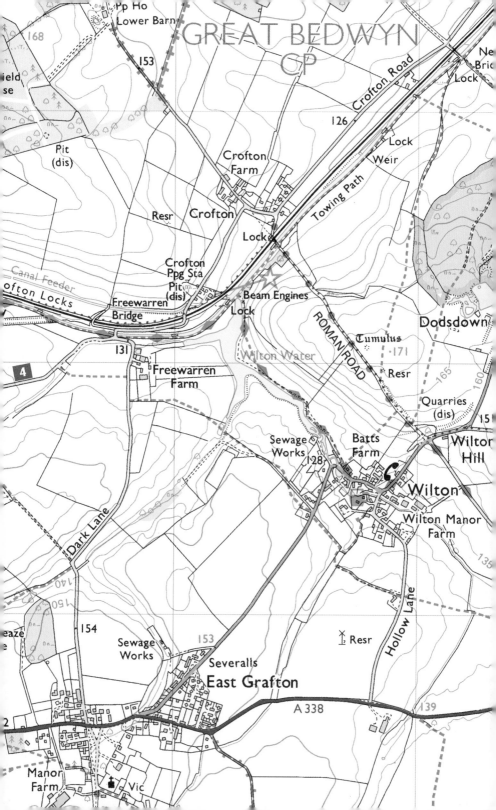

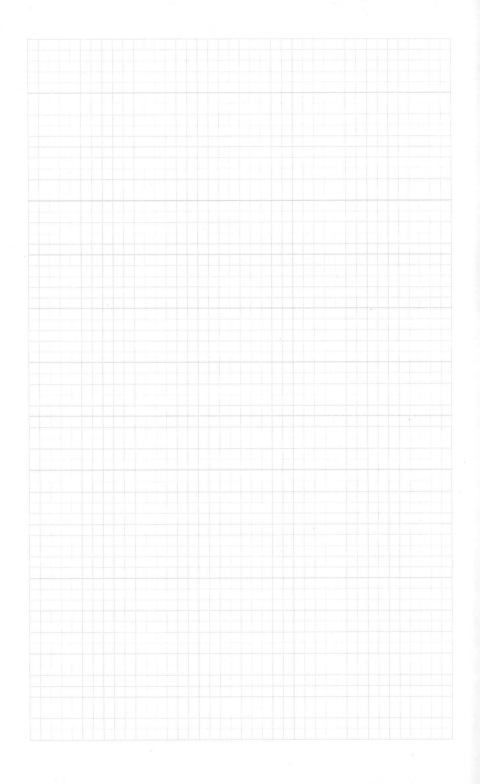

MAP
5

FROM *MARY ROSE*
TO HMS *QUEEN ELIZABETH*

The map: Road Map enlarged to 1:200 000 scale

The story: Dredging up the past

The main channel into Portsmouth Harbour was dredged in 2017 to accommodate the navy's newest and biggest warship, HMS *Queen Elizabeth*. Over three million cubic metres of mud were removed and more than 20,000 items recovered, including eight cannons and a number of cannonballs. A British torpedo, a German sea mine and five large bombs also kept the locals and the Royal Navy's ordnance disposal team on their toes. Rich in defensive fortifications, both on land and at sea, maps tell the ongoing story of Britain's history as a seafaring nation. Four sea forts were built outside the entrance to Portsmouth Harbour and Southampton Water as part of a defensive chain known as the Palmerston Forts. Their contruction was instigated by Prime Minister Henry Temple, 3rd Viscount Palmerston, in response to the 1859 Royal Commission on the Defence of the United Kingdom. By the time they were completed in 1880, the threat of invasion had long since passed. The forts were decommissioned in 1956 – all are now in private hands with two operating as luxury hotels.

■ Easy

1. Can you find a place whose name is the same as 'complain', but with one letter changed?

2. Can you find, on the mainland (the top part of the map), a town starting with 'North' that is further south than a place starting with 'South'?

■ Medium

3. How many museums are marked on the Isle of Wight (the bottom part of the map)?

4. What's the difference in value between the highest-numbered A road on the map and the lowest-numbered A road?

■ Tricky

5. What do the words 'north', 'hedge', 'west', 'worlds' and 'down' all have in common?

6. How many different animals can you find portrayed on map symbols?

■ Challenging

7. If you were at Nettlestone Point looking due east, and turned southwards until you were facing due south, how many campsites would you have faced towards as you turned?

8. Given that the scale of the map is 1:200 000, what is the length of the marked section of Roman road to the nearest kilometre?

MAP
6

HOLDING BACK THE SEA

The map: VectorMap Local at 1:10 000 scale

The story: Thames Barrier, London

The Thames Barrier spans 520 metres and prevents the floodplain of most of Greater London (around 125 square kilometres) from being inundated by exceptionally high tides and dangerous storm surges moving up from the North Sea. It has been operational since 1984 and has been closed (raised) almost 200 times to protect against tidal flooding and for combined tidal/ fluvial events. The barrier consists of ten steel gates with the main gates standing as high as a five-storey building when raised. Each main gate weighs 3,300 tonnes. To allow for maintenance, each can be rotated through 180 degrees out of the river. They can also be rotated to a position that allows the river to under-spill the barrier to control the flow.

Easy

1. How many occurrences of the following words (including words that start with these words) are there on the map?
   a. Mud
   b. War
   c. Thames

2. Can you find two neighbouring parallel streets that start with opposing compass points?

Medium

3. Can you find two roads that are named after animals?

4. Can you find a place whose name starts with a chemical element?

Tricky

5. Can you find two places that are each named after one half of a very well-known married couple?

6. The words 'festoon', 'slip', 'sub' and 'unity' all share a property based on their usage on the map. Can you work out why, and then further divide them into two pairs?

Challenging

7. Given that the scale of the map is 1:10 000, how far is it as the crow flies from the Thames Barrier Centre to the closest war memorial, to the nearest 50 metres?

8. What is the real-life distance between the mean low water line and the mean high water line at the place where they're furthest apart?

COUNTY MAPPING
& NATIONAL GRID

OS Fact: Mapping Mars

After 225 years of mapping Great Britain, and experience of mapping internationally, 2016 saw us take OS mapping out of this world. We created a one-off paper and digital map of the Martian landscape, using NASA open data. We were chosen because our maps are easy to understand and present a compelling visualisation.

MAP
7

WHY WOULD YOU NEED CONTINUOUS MAPS?

The map: County Series 1874

The story: Going around the bend at Barnes

While the triangulation network provided an accurate location across the country, the maps we produced stuck with a county focus that strongly influenced the history and everyday life of Britain at the time. In this example from Surrey (1874), the county of Middlesex is the blank area north of the River Thames. This bend in the river is well known as part of the Oxford and Cambridge Boat Race course. The county names are still used today to describe which side of the river, or station, the boats start from and aim for on the bends and bridges; the Fulham/ Chiswick side is known as Middlesex and the Putney/ Barnes side as Surrey. However, due to boundary changes, Barnes is no longer in Surrey and Middlesex is just a historic remnant.

▨ Easy

1. From the map, would you infer that Hammersmith is north, south, east or west of Barnes?

2. Which one of these four compass points can you find on the map: northeast, southeast, southwest or northwest?

■ Medium

3. The name of which US president, which is also a city in Ohio, appears on the map?

■ Tricky

4. Which location on the map is made up of two female given names, the second of which has had an 'e' removed from the end?

5. The names of which three London tube stations appear on this map?

■ Challenging

6. Which locations on the map could mean the following?
    a. Cleaning substance is effective
    b. Course for bowling
    c. Drop general
    d. Grinder implant

MAP
8

ROOM WITH A VIEW

The map: County Series 1912

The story: Hotel Metropole, Blackpool

The mid-eighteenth-century fashion of travelling to the coast for health and wellbeing led Lawrence Bailey, a local farmer, to start construction of The Metropole Hotel in 1776, right on the edge of the beach at Blackpool. A new private road in 1781 improved access in time for its completion in 1785, making it only one of four hotels in the small but growing resort. The railway arrived in the 1840s, connecting it to the industrial regions of northern England, and Blackpool developed rapidly. In 1855 the tramway was laid down, following the new promenade, but running behind the hotel. This has left the dramatic Metropole standing alone, with great views directly over the beach.

QUESTIONS

▩ Easy

1. Which day of the week appears on the map?

2. Which appears more times on the map: the word 'back', or the word 'bank'? Include both singular and plural versions in your counts.

▩ Medium

3. Which road sounds from its name to be particularly suitable for pedestrians?

4. What two different words on the map are strongly historically associated with 'Victoria'?

▩ Tricky

5. The name of which Oxford college appears on the map?

6. Which words on the map could potentially mean the following?
   a. Lose your footing
   b. Years of a member of the clergy
   c. Meadow of metal coils
   d. Handrail on an underground train

▩ Challenging

7. Which words on the map could solve the following cryptic crossword clues?
   a. Draw back charge
   b. Walk for males with half decade
   c. Confused a celebrant for host storage?
   d. Regular or variable, following currency

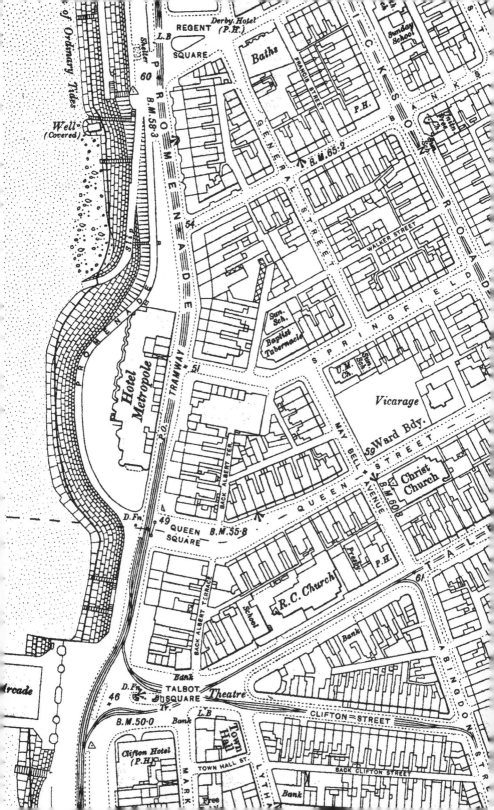

MAP
9

NOT FAKE – JUST A FALSE ORIGIN

The map: Landranger enlarged to 1:40 000 scale

The story: The Central Meridian

In 1936 the Ordnance Survey moved from county-based mapping with their local Cassini projections to a single Transverse Mercator projection covering the whole of Great Britain. This gave us continuous mapping across county and country borders and allowed for the introduction of a 'national grid'. To get a spherical Earth onto flat paper a projection is used. Our 'OSGB36' projection preserves area shapes well and delivers high-accuracy east to west, which is great for our tall, narrow part of the world. The central meridian runs straight up along the 2° west longitude and is seen here coincident with the 00 eastings at Berwick. To avoid negative grid figures, the true origin of the projection was moved to create a 'false' origin for the national grid to the west and south of the Isles of Scilly.

▨ Easy

1. Can you find all of the following on the map?
   a. Eleven places of worship (a black cross, or a black cross with a circle or square beneath)
   b. Six masts (a triangle with a zigzag line on top)
   c. Seven mileposts ('MP')

2. What is the lowest height numerically indicated on the map?

■ Medium

3. What do 'hill', 'prior', 'blue' and 'white' have in common?

4. Which compass point does not appear on the map?

■ Tricky

5. How many parts of the body feature on the map, either in their own right or within other words?

6. Which words on the map sound like they could have the following meanings?
   a. Rescind gratitude
   b. Drill for oil
   c. Pieces of a male sheep
   d. Goat rules

■ Challenging

7. Given that the scale of the map is 1:40 000, how far is the golf course from the camping and caravan site, to the nearest kilometre as the crow flies?

8. Can you find places on the map whose names are anagrams of the following phrases? Ignore the spaces and punctuation, which may differ from those in the place names.
   a. WIN POWERED BUCKET
   b. THROATY SHEEP
   c. ALWAYS SEDUCIVE

SYMBOLS

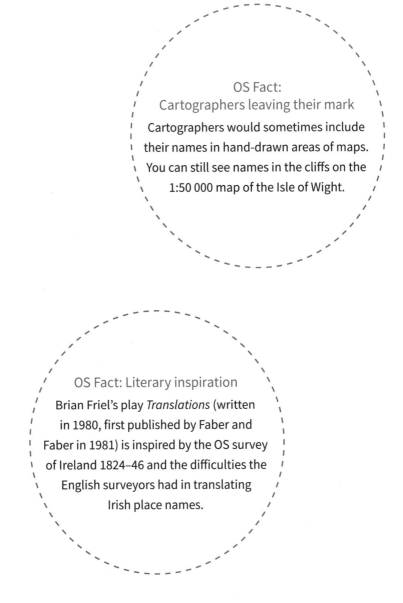

OS Fact:
Cartographers leaving their mark
Cartographers would sometimes include
their names in hand-drawn areas of maps.
You can still see names in the cliffs on the
1:50 000 map of the Isle of Wight.

OS Fact: Literary inspiration
Brian Friel's play *Translations* (written
in 1980, first published by Faber and
Faber in 1981) is inspired by the OS survey
of Ireland 1824–46 and the difficulties the
English surveyors had in translating
Irish place names.

MAP
10

THIS ONE DOESN'T BITE

The map: Explorer enlarged to 1:20 000 scale

The story: Whipsnade White Lion

The 'Whipsnade White Lion' is the largest hill figure in Britain, covering almost two acres and seen here nearly on its back as the hill slopes down to the B4506. It is also the youngest, cut into the chalk hill in 1933 to indicate the location of Whipsnade Zoo. It is difficult to see from the road, but Beacon Hill, near Ivinghoe, off this map to the west, was used during construction to check the accuracy of the outline and offers a good vantage point today. Fading to green with weeds, it has recently been restored thanks to a donation of chalk from a local building project.

■ Easy

1. How many times does the word 'Whipsnade' appear on the map?

2. How many farms are marked on the map?

■ Medium

3. The names of which two colours appear as words on the map?

4. Pair the following words to form places shown on the map: age, bury, dun, hall, ham, high, long, rings, sallow, sew, spoons, spring, stable and stud.

■ Tricky

5. Which is the first letter of the alphabet not to appear capitalised anywhere on the map?

6. If you were driving from Dell Farm to Studhamhall Farm via the only route fully visible on the map, how many left turns would you make?

■ Challenging

7. Given that the scale of the map is 1:20 000, can you calculate the distance between the picnic site and the golf club as the crow flies, to the nearest kilometre?

8. Each of the following phrases is an anagram of two words found on the map, ignoring spaces and punctuation. For example 'Ham frill' would be an anagram of 'Farm' and 'Hill'. Can you identify the two words in each case?
   a. OH NO! SLIME!
   b. ELITE COW DRINK
   c. BRAIN'S TOIL
   d. SWEETER AGE

MAP
11

HAVE A HEART

The map: Explorer enlarged to 1:12 500 scale

The story: Hawthorn Heart Wood

Only just about big enough to meet the specification of our Explorer series, you need to look closely to spot this memorial hawthorn wood in the shape of a heart. Visible from Huish Hill near Oare in Wiltshire, it was created by Lady Keswick, the owner of Oare House, in memory of her two brothers. The 1,000 hawthorn trees were planted in 1999 and produce red blossoms for a short spell each year. Oare House itself was built in 1740 for the London wine merchant Henry Deacon. In the 1890s it had fallen into disrepair but later owners created new formal gardens and extended the house. During the Second World War, it was used by the War Cabinet for secret meetings.

◻ Easy

1. What is the highest number that appears on the map?

2. What place on the map comes first alphabetically?

◼ Medium

3. If you were driving from Pennings Farm (leaving on the southside) to the school in Oare, what's the fewest number of left turns at junctions you could make?

4. What word on the map means 'more or less'?

◼ Tricky

5. Given that the scale of the map is 1:12 500, how far apart are the two wells, as the crow flies, to the nearest 25 metres?

6. If a crow flew directly from Rainscombe House to the nearest place of worship, which farm would it fly over?

◼ Challenging

7. What do the words 'down', 'hill' and 'wilts' have in common?

8. If a crow flew directly from the southernmost non-Roman archaeological site (i.e. location marked in Gothic script) to the westernmost non-Roman archaeological site fully visible on the map, which farm would it fly over?

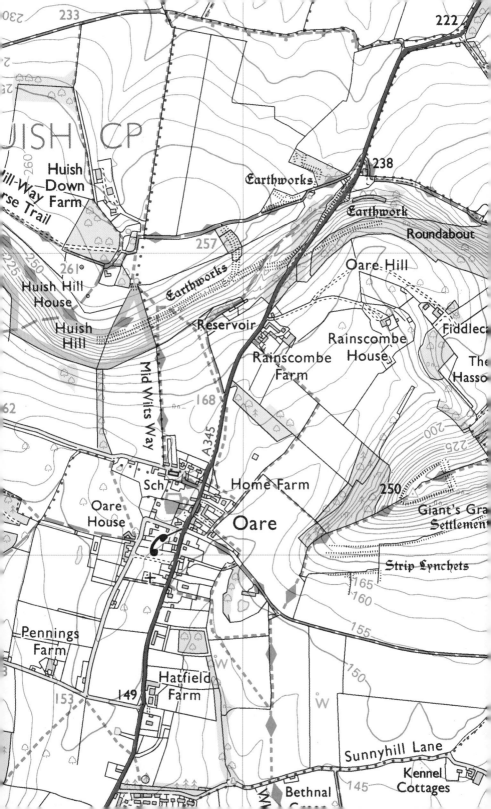

ACCESS FOR ALL

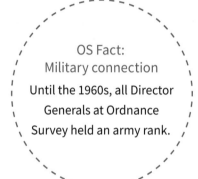

OS Fact:
Military connection

Until the 1960s, all Director
Generals at Ordnance
Survey held an army rank.

MAP
12

THE PROTEST THAT CREATED
THE NATIONAL PARKS

The map: Explorer at normal 1:25 000 scale

The story: Kinder Scout, Peak District

Today, the map of Kinder Scout is highlighted in pale yellow for open access and crossed by the Pennine Way National Trail. Back in the early 1930s, the land was reserved for grouse shooting and was off limits. That was until Benny Rothman and four friends started to ramble there at weekends to escape the smoke and grime of their industrial homes. They were jailed, but this began one of the most successful direct-action protests in British history, with about 10,000 ramblers taking to the hills. Appropriately, the Peak District, described as the 'lungs of the industrial north', was the first National Park to be created, in 1951.

QUESTIONS

■ Easy

1. What is the name of the National Trail, marked with an acorn symbol, that is shown on the map?

2. How many times do the words 'red brook' appear on the map?

■ Medium

3. Which path on the map shares its name with a connection between the Earth and heaven in the biblical book of Genesis?

■ Tricky

4. What connection can you find between 'gates', 'scout', 'downfall' and 'low'?

5. Which locations on the map sound like they might mean the following?
   a. A pig's returned
   b. A sea creature's haunt
   c. Rears of game birds
   d. Parcels of sheep hair

■ Challenging

6. Given that the scale of the map is 1:25 000, how far is Crowden Head from Crowden Tower, to the nearest kilometre as the crow flies?

MAP
13

COMMON LAND

The map: Explorer enlarged to 1:20 000 scale

The story: Space to breathe in West Winch, Norfolk

The legal enclosure of land began in the sixteenth century and removed millions of acres of common land that had been used for feeding livestock and collecting firewood. Enclosure changed the rural landscape of England and played its part in the agricultural and industrial revolutions that followed. Any common land that remained was generally restricted to those with 'commoners' rights'. The Countryside and Rights of Way (CROW) Act 2000 provided 'open access' to these areas for the public to enjoy, and they are shown on Explorer maps in pale yellow. Today, the value of such areas for wildlife and our health and wellbeing cannot be underestimated (and helped, of course, by a map).

◻ Easy

1. How many phone boxes are marked on the map?

2. If you were a farmer, where might you visit in order to sell cows?

◼ Medium

3. How many farms are named on the map?

4. What is the highest number that appears on the map?

◼ Tricky

5. a. What is the fewest number of right turns you could take to drive from the end of one restricted byway to the end of any other restricted byway? Restricted byways are marked by dashed green lines with half crosses.

   b. And what is the fewest number of left turns you could take to drive from the end of one restricted byway to the end of any other restricted byway, without using the same pair of byways already used in your answer to the previous question?

6. Given that the scale of the map is 1:20 000, how far apart are the two places of worship as the crow flies, to the nearest 100 metres?

◼ Challenging

7. How many times do east, west, north and south appear in place names?

8. Can you find places on the map whose names are anagrams of the following phrases? Ignore the spaces and punctuation, which may differ from those in the place names.
   a. REAL QUENCHES
   b. PUN IN YARD
   c. HE GET OLD

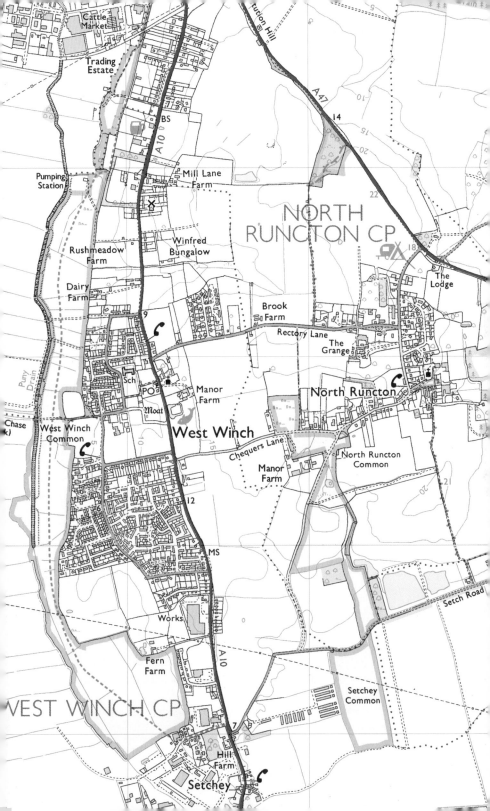

MAP
14

CHECK THE TIDE TIMES

The map: Explorer enlarged to 1:20 000 scale

The story: Steart Island and River Parrett, Somerset

The England Coast Path will become, in 2020, the longest waymarked coastal path in the world. As each section is completed, the OS is updating its Explorer mapping with new 'pink' coastal access areas to reflect the new open access rights along the coast of England. Here, the River Parrett drains the Somerset Levels into the Bristol Channel. The river used to form the boundary between the Anglo-Saxon kingdoms of Wessex and Dumnonia. The dynamic nature of the coast means that the sand, mud and river channels are constantly changing and care is required. Don't forget to check local signage and warnings before you step off the path.

QUESTIONS

▨ Easy

1. How many beacon symbols are shown on the map?

2. Can you find a place name on the map that contains the name of an animal?

▤ Medium

3. How many times does the number '6' appear on the map?

▤ Tricky

4. Which location on the map shares its name with that of a 2005 film directed by Michael Bay?

5. Can you find a location on the map which contains, in order, the letters that make up the word 'sleaze'?

▤ Challenging

6. Can you find places on the map whose names are anagrams of the following phrases? Ignore the spaces and punctuation, which may differ from those in the place names.
    a. SMALL FORD CAR
    b. LEFT STARTS
    c. WELL LEAPT VET

7. Given that the scale of the map is 1:20 000, how far is the place of worship from the parking place as the crow flies, to the nearest 100 metres?

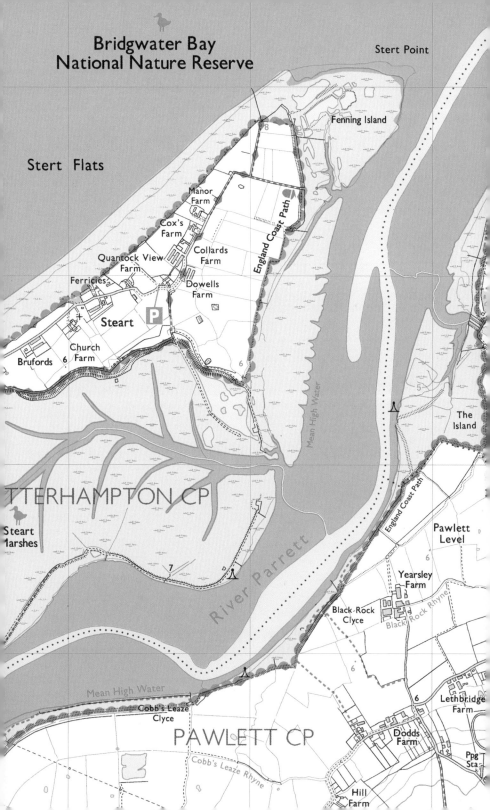

MAP 15

THE LONG AND SHORT OF IT

The map: OS Road at 1:250 000 scale

The story: Pennine Way

The Pennine Way was Britain's first National Trail (see page 69) and was devised by Tom Stephenson in the 1930s. It generally follows the high route along the 'backbone' of England, from the Peak District, up the Pennines to the Tyne Valley and Scottish border. At 256 miles (412km), it is currently the longest trail, and keen ramblers often start in the south and head north to keep the rain at their backs! Cross Fell, at the centre of this map, is the highest point of the trail. Not to be left out, cyclists and horse riders can enjoy the Pennine Bridleway which runs for 205 miles (330km) and explores many of the ancient packhorse routes and drovers' roads across the Pennines.

▨ Easy

1. How many blue train symbols appear on the map?

2. What word can be used to unite Appleby, Brougham and Carrock?

▨ Medium

3. Which place names sound like they could mean each of the following?
   a. A pay increase
   b. A notice of someone at the door
   c. An ursine crossing
   d. A model village
   e. Question bacon

4. What do 'hard' and 'odd' have in common in places on the map?

▨ Tricky

5. What is the highest number of place names ending in 'by' to appear in a single grid square?

6. Which twentieth-century UK prime minister appears on the map?

▨ Challenging

7. Which places on the map could solve the following cryptic crossword clues?
   a. Criticise rap
   b. Cold guy with new, muddled yes for extraction vents
   c. Wife healthy with large mammal

MAP
16

WORTH THE EFFORT

The map: OS Road at 1:250 000 scale

The story: Sandwood Bay, Cape Wrath

Listed among some of Britain's 'most remote spots worth seeing' and described as possibly the most beautiful beach in Britain, Sandwood Bay could well be worth the four-mile hike from the nearest road. Pale white sands, turquoise waters, a shimmering saltwater loch, dunes and a 76 metre rock stack make for a spectacular destination. The bay is exposed to Atlantic breakers and has been the scene of many early shipwrecks, although any remains have long since been washed away or buried under the wide sands. The Cape Wrath lighthouse was built in 1828 and no ship has been wrecked at Sandwood since.

▨ Easy

1. Which is the second-highest peak on the map?

2. Can you find a place that shares its name with a famous physicist?

▧ Medium

3. Can you find a place whose name is one letter different from 'respond'?

4. Looking southwest from the Reay Forest, which is the tallest peak visible on the map?

▨ Tricky

5. What place on the map sounds like a poor telephone connection?

6. And what place on the map sounds like anger caused by a cloak?

▨ Challenging

7. Given that the scale of the map is 1:250 000, how far is the museum in Unapool from the golf course in Balnakeil as the crow flies, to the nearest kilometre?

8. Can you find bold-font place names on the map whose names are anagrams of the following? Ignore the spaces and punctuation, which may differ from those in the place names.
   a. A HEROIC ME
   b. SORT MEEKLY
   c. AN EARLY UNIT

MAP
17

NOT THE HIGHEST
BUT STILL TOPS

The map: Landranger at 1:50 000 scale

The story: Training for Mount Everest

It would be easy to centre a map on just Snowdon, but we've chosen to highlight the Llanberis Pass, between the mountain massifs of Snowdon and Glyderau, which features the popular Tryfan mountain. The hotel at the summit of the pass is now a youth hostel, and the Pen-y-Gwryd Hotel at the eastern end of the pass was the training base for the 1953 Everest expedition. The Cromlech Boulders, on the roadside (under the 'A4086' text on the map) are used for bouldering, a form of rock climbing without ropes. They were saved from a road-widening scheme in 1973 after a six-year protest by local people, climbers, historians and geologists.

Easy

1. How many different English words appear in Gothic text on the map?

2. What is the greatest height labelled on the map with black text? And what is the lowest height in black text?

Medium

3. How many different tourist and leisure information markers (blue symbols) are there on the map?

4. Where on the map might you in some sense expect Lucifer's food to be prepared?

Tricky

5. What feature on the map connects Idwal, Teyrn, Glas and Llydaw?

6. Which (English-language) words or phrases on the map could potentially mean the following?
   a. Upright type faction
   b. Leaders' meeting
   c. Honeyeater's music choice

Challenging

7. Given that the scale of the map is 1:50 000, how far apart are the two stations as the crow flies, to the nearest 250 metres?

8. Can you find places on the map whose names are anagrams of the following phrases? Ignore the spaces, which may differ from those in the place names.
   a. FLY CHARGED
   b. WE RELY ON
   c. WEIRD FLAIR

WHAT'S IN A NAME?

OS Fact: Trig pillars

In 1935 Ordnance Survey began the re-triangulation of Great Britain, to improve the accuracy of our maps and unify mapping on to a single national datum projection and reference system. This led to the OSGB36 datum and the national grid, both of which are still with us today.

The re-triangulation involved erecting around 6,500 concrete pillars (known as trig points) on British hilltops, which were used as reference points.

If you haven't come across it before, triangulation works by determining the location of a point by measuring angles to it from known points at either end of a fixed baseline and, in this case, those known points were the trig pillars. A theodolite would have been secured to the top mounting plate, and angles then measured from the pillar to other surrounding points. But time and technologies have moved on enormously, and the trig pillar's original use is now obsolete.

MAP
18

YOU WON'T FIND ANY BOATS HERE

The map: Landranger enlarged to 1:35 000 scale

The story: Cold Harbour, Lincolnshire

Place names are a remnant of the past, revealing a history that has otherwise been forgotten or lost in the mists of time. Cold Harbour is thought to refer to abandoned settlements or dwellings used as refuges by travellers in Anglo-Saxon times. A handful of Cold Harbour names remain around the country as modern place names. This one near Grantham certainly supports the idea, with the important London to York 'Old North Road' running through it. While the Roman name for this road is not known, it is called Ermin Street ('Earninga Straete' in Old English) after the Earningas tribe.

Easy

1. Which place name on the map includes the name of a major capital city?

2. What is the highest number that appears on the map?

Medium

3. Which mast, indicated by a triangle with a zigzag on top, is furthest from a marked phone box?

4. Which of the following would you expect to be able to do in this map region?
   a. Play golf
   b. Visit a castle
   c. Visit a National Trust site

Tricky

5. Which is the closest location to Cold Harbour that is labelled with something designed to contain water?

6. And what is the second-closest location to Cold Harbour that is labelled with something designed to contain water?

Challenging

7. Given that the scale of the map is 1:35 000, which of the farms shown on the map is approximately 3 kilometres from Cold Harbour?

8. Which location sounds like it could be an instruction to fasten your seat restraint after stopping your car?

MAP
19

KWIRKAPIDDLE

The map: Landranger enlarged to 1:40 000 scale with aerial imagery

The story: Foula, Shetland Islands

Foula – 'Bird Island'. The name derives from two Old Norse words, for bird (fugl) and island (ey). There are many islands in the west of Britain and around Scotland with an 'ey' ending (now written 'ay') – Rousay, Sanday, Westray in the Orkneys, for example. Due to the remoteness of Foula, the island has retained many more old names scattered across a landscape that would be easily recognised by the Vikings. Can you also spot what looks wrong with this map? Usually, when we add hill shading, the shadows are cast from the top left corner. Here we've overlaid real-world aerial imagery showing where the sun really shines – from the south.

Easy

1. How many occurrences of the word 'cave' are there on the map, either in the singular or the plural?

2. Can you find places whose names fit the following patterns, where an underline represents a single missing letter?
   a. S_I_N_W_L_E
   b. C_D_A_I_L_
   c. H_D_I_L_V

Medium

3. How many occurrences of 'ham' are there on the map, both on its own and within other words?

4. Can you find two places which contain (in one place) the name of a farmyard animal and (in the other place) the sound it makes?

Tricky

5. Which place name shares its name with a type of bird?

6. Which locations on the map sound like the following?
   a. Calls the principal
   b. Plait match

Challenging

7. Which location contains the name of a character from the long-running sitcom *Friends*?

8. How many occurrences of 'da' are there on the map, both on its own and within other words?

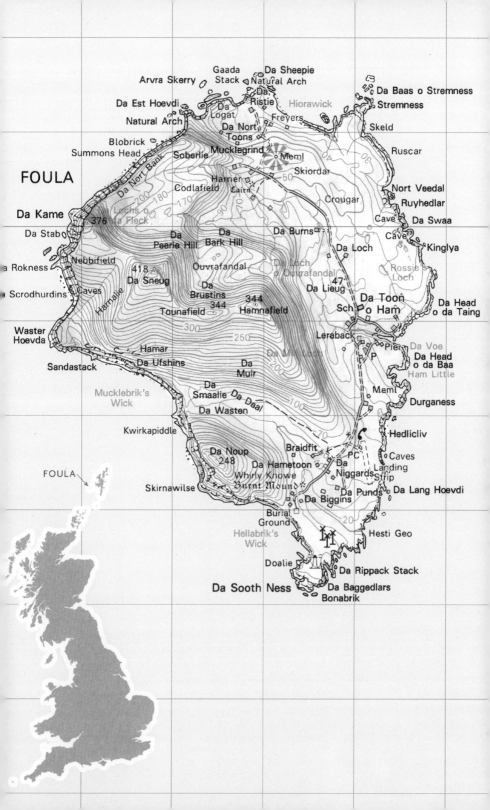

CHANGING LANDSCAPES

OS Fact: Benchmarks

While trig pillars were used as accurate fixed points for mapping coordinates in the national grid, benchmarks were fixed points that were used to calculate a height above the mean sea level, and have been around for much longer than trig pillars.

The horizontal marks are used to support a stable 'bench' for a levelling stave to rest on – hence 'benchmarks'. This design ensured that a stave could be accurately repositioned in the future and that all marks were uniform.

Some benchmarks in our archive date back to 1831. The last traditionally cut arrow-style benchmark was carved on a milestone outside the Fountain pub in Loughton in 1993.

More than 500,000 benchmarks were created, but their number is reducing as roads change and buildings are demolished.

MAP
20

NOT SO HAPPISBURGH

The map: Explorer at 1:25 000 scale with old coastline overlay, 1905

The story: Mapping erosion at Happisburgh, Norfolk

With some of the fastest-eroding cliffs in the country, Happisburgh highlights the value of mapping over time. In a game of winners and losers, this stretch of sandy coastline is being allowed to erode naturally. The map has been overdrawn with the high and low water marks from old OS six inch mapping. The beach has become a thin thread of sand compared to a hundred years ago, with the coastline south of Happisburgh losing the most ground to the sea. They certainly thought ahead when the Happisburgh lighthouse was built in 1790. Maybe in a hundred years the volunteers who run the oldest working lighthouse in East Anglia will be able to open a beachfront café.

■ Easy

1. How many wells, marked with a blue 'W', are there on the map?

2. How many occurrences of the word 'green' are there on the map?

■ Medium

3. How many different types of tree appear as words on the map?

4. Which road has a name that it implies it covers only a brief distance?

■ Tricky

5. What is the connection between church, lighthouse, orchard, moat, chimney, hall and college?

6. If you were driving from Brunstead House Farm to Manor Farm, what is the lowest number of right turns at junctions you could make? Ignore junctions with tracks.

■ Challenging

7. By the sound of its name, where could you go for metalwork?

8. What street on the map could be read to mean that the crowning of a new king or queen is soon to occur?

NORTH SEA

Low Water Mark of
Ordinary Tides (1905)

High Water Mark of
Ordinary Tides (1905)

Sand (1905)

1905 coastline from OS six inch
mapping redrawn over current
1: 25 000 scale Explorer

Mean Low Water
(2017)

Mean High Water
(2017)

Happisburgh

England Coast Path

HAPPISBURGH
CP

LESSINGHAM
CP

Lessingham

MAP
21

MAN-MADE SCARS

The map: County Series

The story: Greengates Quarry, Teesdale

The landscape, soils and underlying geology dictate much of what we do, what we build and what we use the land for. Stone has always been a building material of choice, but not easily wrested from the ground. This quarry at Greengates reveals the dramatic history of the rocks here in the North Pennines. At the end of the Carboniferous period, about 300 million years ago, this area in Lunedale suffered tilting and faulting, with land on the southern side displaced downwards. This allowed dolerite, a molten rock, to come up between the beds of limestone and sandstone. It is an ideal roadstone, and the quarry here highlights the shape and size of one such intrusion, now worked out.

■ Easy

1. How many times does the word 'quarry' appear on the map?

2. How many times does the abbreviation 'F.P.' appear on the map, indicating a footpath?

■ Medium

3. How many different words appear on the map?

4. What is the highest number that appears on the map? For the purposes of this question, consider that the numbers each side of a dot are two separate numbers.

■ Tricky

5. What do the words 'ill', 'lack', 'rough' and 'rushing' have in common?

6. Can you find a colour of the rainbow and a shade of that colour on the map?

■ Challenging

7. If you draw a path that joins the numbers from 304 to 310 in sequence, what letter of the alphabet most closely resembles the resulting path?

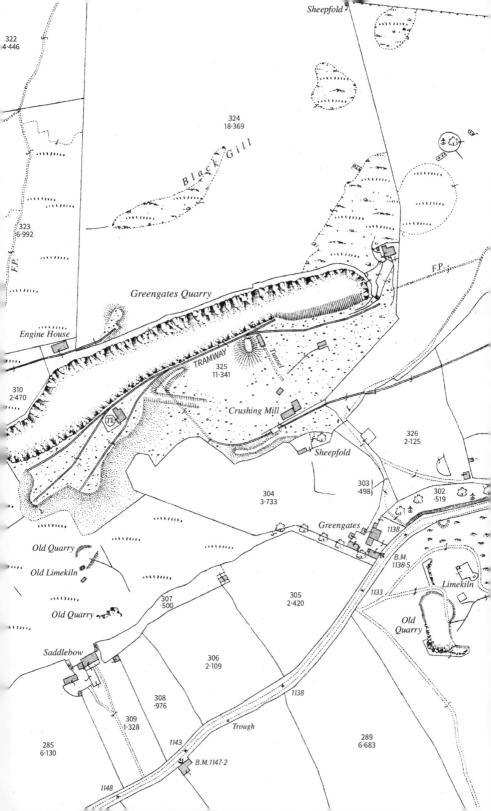

MAP
22

TRAINS CANCELLED
DUE TO FLOODING

The map: 1 inch (1963) at 1:40 000 scale with overlay

The story: Kielder Water

By the time of this map, this section of the Border Counties Railway had been closed to passengers for seven years. Plashetts station, despite its remote location, was quite substantial and served a waggonway branch up an incline to Plashetts Colliery. In the late 1960s plans for a large reservoir to support a booming industrial economy saw the construction of Kielder Water. Changing fortunes looked to make the project a white elephant even before it was finished. Holding back 200 billion litres of water, it is the largest artificial reservoir in the UK (Rutland Water has the largest surface area). Along with Kielder Forest, the biggest man-made forest in Europe, the area now has important environmental and recreational value. The dark area on the map is an overlay of the reservoir showing what was lost.

▨ Easy

1. How many earthworks are marked on the map?

2. What is the lowest elevation number that appears on the map?

▨ Medium

3. Which word on the map shares its name with a car manufacturer, and how many times does it appear?

4. How many occurrences of the word 'burn' are there on the map?

▨ Tricky

5. Which three places on the map contain within their names an imperial unit of weight?

6. Which four words on the map begin with the name of an animal?

▨ Challenging

7. What place shares its name with a central London Underground station once its first letter has been removed?

8. Which location on the map might be said to have a 'long arm'?

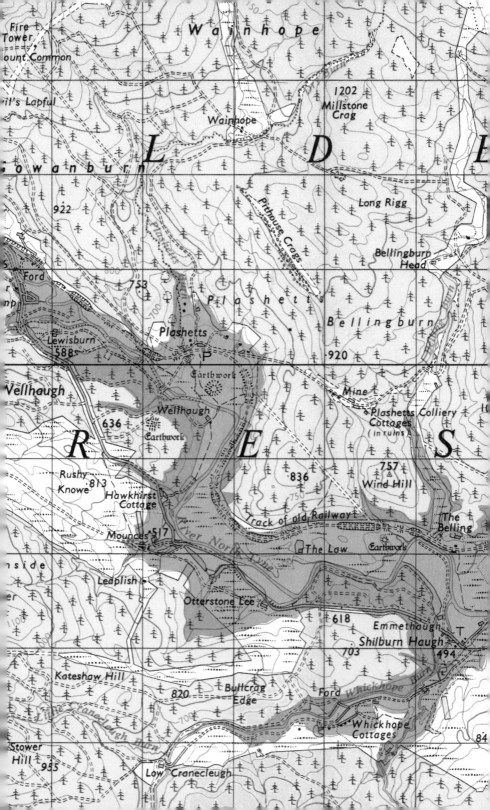

MAP
23

FAMOUS FOR CONCRETE COWS

The map: 1 inch (1946) and Landranger combined at 1:50 000 scale

The story: Milton Keynes New Town

Designated as a new town on 23 January 1967, Milton Keynes was designed, from the outset, to become a city in scale. Three existing towns and fifteen villages were subsumed into it, including the little village it was named after. The 'New City' had a target population of 250,000, and the site was deliberately chosen to be equidistant from London, Birmingham, Leicester, Oxford and Cambridge, with the goal of becoming self-sustaining rather than a London overspill town. The original rule of 'no building higher than a tree' has been relaxed in recent years as Milton Keynes matures into a modern city with its own identity. This map combines the OS 1946 1 inch 'New Popular Edition' with a slice of current Landranger mapping showing the dramatic changes this once rural landscape has undergone.

QUESTIONS

Easy

1. How many words on the map start with 'Wave'?

2. What is the highest number that appears on the map?

Medium

3. What place name combines the name of a famous poet with the name of a famous economist?

4. Which location on the map would appear last if they were all listed alphabetically?

Tricky

5. Which two locations share their names with families from long-running, eponymous US television shows? And which third location shares its name with the location of one of these television shows?

6. Which locations on the map sound like the following?
   a. Angry finish
   b. Small pile of building blocks
   c. Place for biting insects to cross a river
   d. Good skill with a salad tool

Challenging

7. In which location on the map might you in some sense expect to find a partridge?

8. Given that the scale of the map is 1:50 000, can you find two locations that are approximately 1 kilometre apart as the crow flies and which form antonyms of one another?

MAP
24

WHEN RETAIL THERAPY STRIKES

The map: Explorer enlarged to 1:20 000 scale

The story: Lakeside Retail Park, West Thurrock

With rising living standards and greater car ownership, the out-of-town retail centre has come to dominate our shopping and leisure habits over the last 30 years. Lakeside in West Thurrock is not only a great study for the geographer, but has grown to be the largest retail area in Europe and a tourist attraction in its own right. Despite being a former chalk pit and located near industrial areas, good communication links have encouraged the growth of housing, making it a desirable place to live too. So maybe it isn't quite as out-of-town as it was! Can you spot what is different about the style of this map to the usual Explorer?

Easy

1. The name of which planet appears on the map, either in its own right or within another word?

2. Which two words on the map are direct opposites of one another?

Medium

3. Which letter of the alphabet is the only one that doesn't appear on this map? And what is the first letter of the alphabet, working from A to Z, that doesn't appear on the map capitalised?

4. If you were to draw a path joining all of the schools marked on the map, reading from west to east, what letter would this most closely resemble?

Tricky

5. In how many different colours do numbers appear on the map? Count both shades of orange as the same colour.

6. Pair up the following words, to form either words or places marked on the map: an, barn, brick, broom, cause, hill, lake, me, side, and way.

Challenging

7. Given that the scale of the map is 1:20 000, how far apart are the two most distant places of worship, to the nearest 250 metres as the crow flies?

8. Which words on the map could solve the following cryptic crossword clues?
   a. Racing leader bottles drawing, perhaps
   b. Two vehicles join around a group of people?
   c. Real and imaginary group of buildings

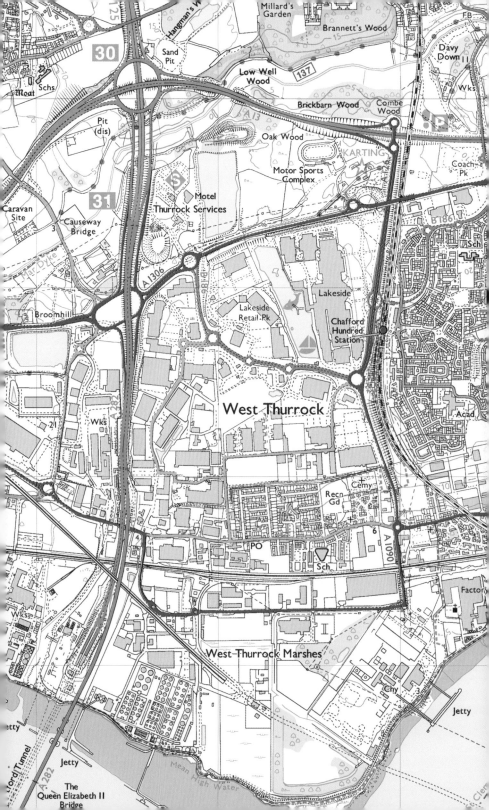

MAP
25

ALL BOXED UP

The map: 1 inch (1948) enlarged to 1:50 000 scale with overprint

The story: Port of Felixstowe

This 1948 map of the small dock at Felixstowe was published just a few years before the revolution brought about by the development of the shipping container. The 'box' did away with the need for a large workforce manhandling thousands of loose goods at risk from damage and theft. No longer did the producers or users of these goods need to be close to a port and, as a result, those in cities like London and Liverpool went into decline. Felixstowe, with its deep water channel close to the sea and acres of available space, was ideally placed to take advantage and became the country's first operational container port. It now handles over 40 per cent of Britain's containerised trade.

▨ Easy

1. How many times does the word 'pier' (or 'piers') appear on the map?

2. How many lighthouses are pictured on the map?

▨ Medium

3. The names of which two classical elements appear on the map?

4. Which two locations have names that begin with colours? And which two locations have names that begin with birds?

▨ Tricky

5. The (pen) name of which twentieth-century author appears on the map?

6. What do the words 'hill', 'croft', 'manor' and 'walk' have in common?

▨ Challenging

7. If you were to draw a path joining all of the instances of the word 'tower', reading from west to east, what letter would this most closely resemble?

8. Can you find names on the map which are anagrams of the following phrases? Ignore the spaces and punctuation, which may differ from those on the map.
   a. RIG A CAVE
   b. HURRAH! I ARCH BOW
   c. SYMMETRY TRIAL

MAP
26

DON'T MISS YOUR EXIT

The map: Vector Map Local at 1:10 000 scale

The story: Gravelly Hill M6 Interchange (Spaghetti Junction)

The first and original 'spaghetti junction' on the M6 at Birmingham was so nicknamed in an article about the new motorway in the *Birmingham Evening Mail* on 1 June 1965. The junction only covers 1 kilometre of the M6 motorway but has 4km of slip roads. The map only hints at the complexity of roads on five different levels with 559 concrete columns, some reaching a height of 24m. Over 21km of carriageway were elevated to accommodate the existing canal, river and railway lines. Columns that came close to the canal had to be positioned so that a horse could still use the towpath beside the water.

▨ Easy

1. Which road shares its name with a famous London recording studio?

2. Which park neighbours an area with multiple streets named after ducks?

■ Medium

3. How many times does the number '6' appear on the map as an individual digit (i.e. not as part of a larger number)?

4. Which word completes all three of these street names: Place, Road and Approach?

■ Tricky

5. Which road shares its name with a Monopoly board property?

6. Which locations on the map sound like the following?
   a. Domesticated waterway
   b. Path over a noisy bird

■ Challenging

7. Which location on the map shares its name with a 1978 UK number one single?

8. Which roads on the map share their names with:
   a. A four-term UK prime minister?
   b. A well-known movie spy?

MAP
27

A GOLDEN MOMENT IN TIME

The map: Vector Map Local 2012 at 1:10 000 scale

The story: Queen Elizabeth Olympic Park

Paper maps and their digital equivalent, the image tile, are curated products. A snapshot in time. This map of Stratford in east London, the location of the Olympic Park for the London 2012 Games, depicts the area with all the temporary buildings and facilities put in place for that moment in time. Afterwards our maps were updated again as buildings not required as part of the Olympic legacy were torn down or remodelled. Of course, this happens with all our maps but normally over much longer timescales, with the familiar national grid providing that friendly link across time.

Easy

1. How many masts are marked on the map?

2. How many stadiums and arenas are marked on the map?

Medium

3. Which entry on the map provides two possible spellings, both of which are valid?

4. What do 'stadium', 'village' and 'aquatic' all have in common?

Tricky

5. Which locations on the map sound like the following?
   a. Street with a larger stick
   b. Place where tall fermented honey drinks meet
   c. The pepper containers of a place of worship
   d. A European country, trademarked
   e. Three different types of water
   f. A significant street

6. Given that the scale of the map is 1:10 000, how far is the statue from the nearest mast? Give your distance as the crow flies, to the nearest 25 metres.

Challenging

7. Which Olympic venue contains a hidden European capital city?

8. Can you find places on the map whose names are anagrams of the following phrases? Ignore the spaces and punctuation, which may differ from those in the place names.
   a. SORT DRAFT
   b. KNELL IN INTERNAL LAUNCH
   c. RUNS THREATEN FOLLOWER

INDUSTRIAL HERITAGE

OS Fact: Dracula

Bram Stoker was a fan of our work.
Here's a quote from the first chapter of
Dracula: 'I was not able to light on any map
or work giving the exact locality of the Castle
Dracula, as there are no maps of this country
as yet to compare with our own Ordnance
Survey maps; but I found that Bistritz, the
post town named by Count Dracula, is
a fairly well-known place.'

BEST LAID PLANS

The map: Explorer at 1:25 000 scale with overprint

The story: Somerset and Dorset Canal

A plan for a major canal, the Somerset and Dorset, was initiated in 1792 to link Bristol with Poole. A branch of the canal from Nettlebridge to Frome was started first to serve the Somerset coalfield. The canal employed William Smith to survey the local geology of the route. His studies into the rock strata, begun in a nearby mine and furthered by his work on the canal, eventually led to him being described as the 'father of English geology'. By 1803 problems with cost brought construction to an early end and the canal was never completed. The course of the branch canal is shown here, dashed in blue. Only a few remnants survive, including the curved earthworks shown on the map at Vobster Cross.

Easy

1. What is the greatest height on the map?

2. What appears more frequently: 'Vobster' or 'Mells'?

Medium

3. What do the words 'branch', 'park' and 'prospect' have in common?

4. And what do the words 'column', 'duckery', 'gorse' and 'hare' have in common?

Tricky

5. Which two words on the map (including words found at the start of other words) are direct opposites of one another?

6. Which locations on the map sound likethey could mean the following?
   a. Estate of icy tubs
   b. Digit-creation facility
   c. Place where ducks are shaved

Challenging

7. Given that the scale of the map is 1:25 000, what is the distance between the two most far apart milestones (marked 'MS') as the crow flies, to the nearest half-kilometre?

8. Which places on the map could solve the following cryptic crossword clues?
   a. Keen on submerging broken ship for swimmers?
   b. Throw jams in dens
   c. Woman's guy with a weird tally confused previous tracks?

MAP 29

ALL STEAMED UP

The map: OS 6 inch (1901) redrawn and coloured

The story: Great Western Railway, Swindon Works

When the Great Western Railway arrived at Swindon in 1840 it was a quiet little market town with a population of less than 3,000. The GWR were already looking to build all their own railway engines and Swindon fitted the bill. With a canal to the Somerset coalfield, the new junction for the branch line to Gloucester and the need to service and change engines to handle the hills to the west, Swindon offered the ideal greenfield location. Once established, the GWR Works grew to a manufacturing hub with over 14,000 workers that could produce almost every item the railway needed.

QUESTIONS

▦ Easy

1. This is a map about the Great Western Railway, but which two cardinal points feature on the map?

2. What is the highest number that appears on the map?

■ Medium

3. What do 'bridge', 'fleet' and 'high' all have in common?

4. How many times do each of the following abbreviations appear on the map?
   a. S.B.
   b. S.P.
   c. F.P.
   d. B.M.

■ Tricky

5. How many streets on the map are named after UK cities?

6. Which two streets share their name with a Tudor king or a Tudor queen of England?

■ Challenging

7. Using the scale on the map, calculate the maximum possible distance you could travel in a straight line from one side of the Recreation Ground to the other, to the nearest 100 metres.

8. Which places on the map could solve the following cryptic crossword clues?
   a. Practice passage for infantry tested here?
   b. Insect climbing over broken hock for flow controller
   c. Motor moults for trains?
   d. Brother kicks around over back line to block factory

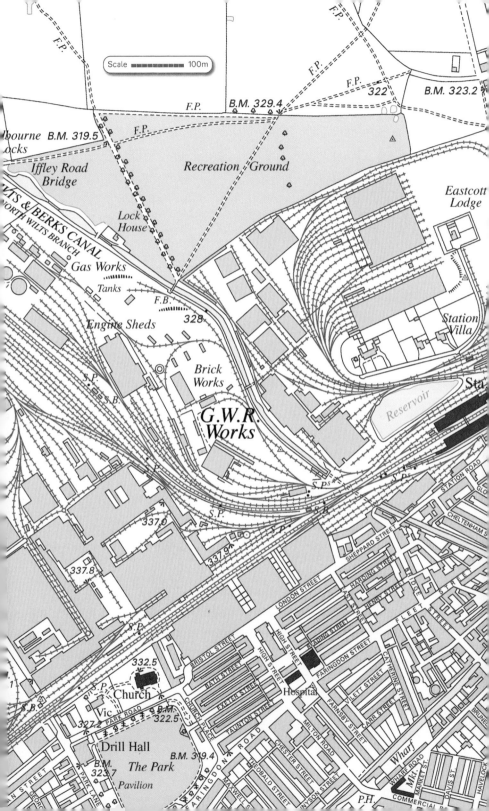

Scale ▬▬▬▬▬▬▬ 100m

F.P.

F.P.

F.P.

F.P.

F.P.

322

B.M. 323.2

B.M. 329.4

F.P.

bourne B.M. 319.5
ocks

*Iffley Road
Bridge*

Recreation Ground

△

*Eastcott
Lodge*

TS & BERKS CANAL

NORTH WILTS BRANCH

*Lock
House*

Gas Works

Tanks

F.B.

328

Engine Sheds

*Station
Villa*

*Brick
Works*

S.P.

S.B.

*G.W.R
Works*

Reservoir

Sta

△

S.Ps

S.Ps

S.B.

337.0

S.P.

337.9

S.P.

STATION ROAD

GLO

CHELTENHAM

337.8

SHEPPARD STREET

HARDING STREET

FLEET STREET

HENRY STREET

LONDON STREET

S.P.

READING STREET

332.5

BRISTOL STREET

BATH ROAD

HIGH STREET

HIGH STREET

FARINGDON STREET

CATHERINE STREET

GILLETT STREET

CARR STREET

Church

Vic

S.P.

PARK ROAD

B.M.

322.5

Hospital

FARNSBY STREET

327.2

EXETER STREET

TAUNTON STREET

NYSON STREET

MILTON ROAD

CHESTER STREET

HAVELO

BRUN

Drill Hall

B.M. 319.4

The Park

FARINGDON ROAD

THEOBALD STREET

Wharf

WHARF ROAD

MARKET ST.

DAVIS ST.

B.M.

323.7

Pavilion

COMMERCIAL

P.H.

PARK LANE

MAXWELLS

MAP
30

RESTORING THE LANDSCAPE WITH ART

The map: Explorer enlarged to 1:20 000 scale

The story: Sultan Pit Pony Earthworks, Penallta, Wales

Can you spot the prancing horse in this map? Some 60,000 tons of coal shale rock were heaped into the shape of a leaping horse by artist Mick Petts. Created as a windbreak for the show arena in Penallta Country Park, the earthwork honours the pit ponies that were used to haul mineral wagons in the underground coal mines, the last of which retired in 1999. The community here nicknamed it 'Sultan' after a particularly famous local pit pony. Because it was created to be a hill, it is best appreciated from the air, unlike the ancient white horses cut into chalk ridges that stride across southern Britain.

▧ Easy

1. How many footbridges ('FB', or 'FBs') are marked on the map?

2. How many 'Park's can you find on the map?

■ Medium

3. What are the lowest and the highest numbers marked on the map?

4. How many different types of public access route – dotted or dashed routes marked in green or orange – are marked on the map?

■ Tricky

5. Can you find the following within the names of locations on the map? They can appear either as a word in their own right or within another word.
   a. A colour of the rainbow
   b. Another colour of the rainbow
   c. The name of a country

6. How many words on the map begin with 'Pen'?

■ Challenging

7. Given that the scale of the map is 1:20 000, how far as the crow flies is the furthest north place of worship to the furthest south place of worship, to the nearest kilometre?

8. Which words on the map suggest each of the following?
   a. Part of a teapot
   b. A court hearing
   c. A 1970s film

MAP
31

DON'T LOOK OVER THE SIDE

The map: Explorer enlarged to 1:20 000 scale

The story: Pontcysyllte Aqueduct

Now a marvel of the Industrial Revolution and a World Heritage Site, the Pontcysyllte aqueduct on the Ellesmere canal strides across the steep-sided Llangollen valley 38m above the River Dee. This more difficult route was chosen to take it through the mineral-rich coalfields of North East Wales. It took ten years to complete and was opened in November 1805. Despite having only one previous and more modest example to his name, and amid public scepticism and concerns about cost, Thomas Telford changed the original plans from a series of locks and an embankment over the river to an uninterrupted 'stream across the sky'.

▨ Easy

1. How many times do the following features appear on the map?
   a. Foot bridges (marked by 'FB')
   b. Wells (marked by a blue 'W')
   c. Public phones (marked by a phone symbol)

2. What is the height difference between the lowest and highest elevations marked on the map?

▨ Medium

3. How many times does 'Fron' appear on the map, either as a word in its own right or as the start of another word?

4. How many different types of body of water are indicated by blue lettering on the map?

▨ Tricky

5. Which farm contains within its name the title of an Oscar-winning film?

6. How many different blue symbols for tourist and leisure information can be found on the map?

▨ Challenging

7. Given that the scale of the map is 1:20 000, what is the distance between the centres of the two Wern Woods, to the nearest 500 metres as the crow flies?

8. If you were to trace a path from the disused pit to the aqueduct, and then on to the cemetery, to the pair of schools and on to the solar farm, what letter would you have traced the shape of?

HISTORY PINPOINTED

OS Fact:
Where is the geographic
centre of Great Britain?

The truth is, there can be no absolute centre for a three-dimensional land mass sitting on the surface of a sphere and surrounded by the ebb and flow of sea water. If you consider the movement of tides on a beach, the shape of the object will change on a constant basis.

We calculate the centre of Great Britain using the gravitational method. This calculates the point at which the object would balance horizontally on the head of a theoretical pin – its centre of gravity. The gravitational method has been used as a scientific application by everyone from Captain Cook to NASA.

We take Great Britain as the mainland and the 401 associated islands shown at the 1:625 000 scale of mapping. We made the calculation by linking our 1:625 000 database with a computer program based on the gravitational method.

On this basis, the centre of Great Britain is a location 7km northwest of Dunsop Bridge, Lancashire, by Whitendale Hanging Stones on Brennard Farm in the Forest of Bowland (SD 64188.3, 56541.43).

WHAT?
NO TRAMLINES!

The map: 1 inch (7th Series 1958) enlarged to 1:30 000 scale

The story: Improving road safety

Many inventions happen as a result of the seemingly unrelated or unexpected coming together at the same time. Of course, it takes someone to see it, or maybe, in this case, not see it! Driving home on a dark and foggy night in 1933 from Queensbury to Booth Town, Percy Shaw had been using the tramline reflections from his headlights to guide him … until they disappeared. In the darkness he spotted two bright points – the eyes of a cat. From this came the idea of reflective road studs, or 'cat's eyes', leading to improved road safety. By the time of Percy's journey, the Halifax tram system was in decline and was closed by 1939.

Easy

1. How many times does 'Hill' occur on the map?

2. How many times is the word 'tunnel' marked on the map?

Medium

3. Which location on the map shares its name with a film that won three Oscars at the 2018 Academy Awards?

4. What is the highest number of places of worship that appears in a single grid square (including those that are only partially within the square)?

Tricky

5. Can you form the words 'beggar', 'ill', 'ton' and 'worth' into two pairs according to a common rule?

6. Which locations on the map sound like the following?
    a. Lair of furnaces
    b. Roof of a financial establishment
    c. Royals hide things underground
    d. Bears something of value

Challenging

7. Given that the scale of the map is 1:30 000, how close together are the two closest places of worship that also have a tower (marked by a rectangle with a cross)? Give your distance as the crow flies, to the nearest 100 metres.

8. Can you find places on the map whose names are anagrams of the following phrases? Ignore the spaces and punctuation.
    a. BARREL MONTH
    b. A KITCHEN RASCAL
    c. STORM'S CUSP

MAP
33

FROM RUTS TO POT-HOLES

The map: Vector Map Local at 1:10 000 scale

The story: Radcliffe Road, Nottingham

This unassuming stretch of road in Nottingham, Radcliffe Road, was the first tarmac road in the world. Tarmac was invented after an accidental spillage of tar was covered with waste slag from a nearby furnace by locals attempting to deal with the sticky mess. County surveyor for Nottinghamshire, Edgar Hooley, happened to be walking in Denby in 1901 and came across the spillage site. Being in a position to develop and perfect a heated mix of tar, slag and stone chips, he began transforming road surfaces, starting with a five-mile stretch of Radcliffe Road. In 1903 he registered 'Tarmac' as a trademark and formed Tar Macadam Syndicate Ltd. Unfortunately, he wasn't any good at selling the stuff and by 1905 it had been bought by a steelworks owner who could make profitable use of his waste slag, creating Tarmac Ltd.

▨ Easy

1. How many places of worship ('PW') are marked on the map?

2. How many times does the word 'meadow' (or 'meadows') appear on the map?

▨ Medium

3. The names of how many different sports appear on the map?

4. How many streets share their name with former monarchs of the UK or England?

▨ Tricky

5. Which locations on the map could be read to potentially mean the following?
   a. Decline a furnace
   b. Become liquid in the street
   c. Way to grind one coin

6. Which two roads share their names with capital cities?

▨ Challenging

7. Given that the scale of the map is 1:10 000, what is the distance between the two furthest apart pavilions as the crow flies, to the nearest 100 metres?

8. Can you find places on the map whose names are anagrams of the following phrases? Ignore the spaces and punctuation, which may differ from those in the place names.
   a. RUGGED CONCERT? DRINK BITTER!
   b. CONVERTING AN ENEMY
   c. JUST IN: NO INNOCENT

MAP
34

ECHOES OF A COMMUNITY

The map: Explorer at 1:25 000 scale

The story: Gainsthorpe Medieval Village, Lincolnshire

It is often difficult to uncover the reason for the desertion of a medieval village. Most likely it would have been a combination of factors that prevented it from remaining a viable community. Plague, soil erosion and deliberate depopulation by landlords for pasture are popular explanations. The text 'site of' on the map informs us that not much is left of the settlement of Gainsthorpe in Lincolnshire, one of about 3,000 that we know of and one of the more clearly defined and best preserved. It was well placed on the major communications routes of the day – the ancient ridgeway (now the B1398) to the west and Roman road of Ermine Street to the east. Did these contribute to its downfall?

▨ Easy

1. How many times does the word 'quarry' (or its plural, 'quarries') appear on the map?

2. How many different symbols for places of worship are there on the map?

▨ Medium

3. What do the words 'cliff', 'old', 'home', 'gain', 'quarry', 'north' and 'spring' all have in common?

4. How many times do the following words appear on the map, either as a word in their own right or within another word?
   a. Cliff
   b. Kirton
   c. Gainsthorpe

▨ Tricky

5. Can you find two words on the map which are also the names of animals?

6. What is the numerical difference between the highest and the lowest numbers on the map?

▨ Challenging

7. Given that the scale of the map is 1:25 000, how far apart are the two public houses as the crow flies, to the nearest kilometre?

8. Which places on the map could solve the following cryptic crossword clues?
   a. Functions right within pans
   b. Recent danger damaged plant store

MAP
35

SPACE AGE RELIC
STILL DELIVERING

The map: Explorer enlarged to 1:12 500 scale

The story: Goonhilly Earth Station, the Lizard, Cornwall

The satellite ground station at Goonhilly was located here, partly because the Lizard is one of the most southerly points on mainland Britain, but also to tie into the underwater communication cables, laid before the space age, that come ashore nearby. Twenty-three of the dish antennas are currently active, including No. 1 'Arthur'. Built in 1962, it was used to carry the first ever transatlantic pictures from the USA to Europe via the Telstar satellite. Soon to become part of Jodrell Bank's radio telescope network, 'Arthur' is 1,100 tonnes of concrete and steel that still has a tracking accuracy of 1/100th of a degree. Nearby is a Neolithic standing stone, so people have been looking up from here for a very long time.

◻ Easy

1. How many times do either 'Tumulus' or 'Tumuli' appear on the map?

2. What is the highest number marked on the map, ignoring the B road?

◼ Medium

3. How many different types of body of water are labelled with text on the map?

4. Which place name on the map would appear first alphabetically?

◼ Tricky

5. How many times does the letter 'b' appear on the map?

6. How many occurrences of double letters are there on the map?

◼ Challenging

7. Given that the scale of the map is 1:12 500, what is the greatest distance between two masts on the map as the crow flies, to the nearest 100 metres?

8. Each of the following phrases is an anagram of two words found on the map, ignoring spaces and punctuation. For example 'YOUTH CLUB'S MUM' would be an anagram of 'Tumulus' and 'Bochym'. Can you identify the two words in each case?
   a. ACTOR FITS? NOT!
   b. RULER OF SECOND
   c. SANG. DIDN'T SNOW

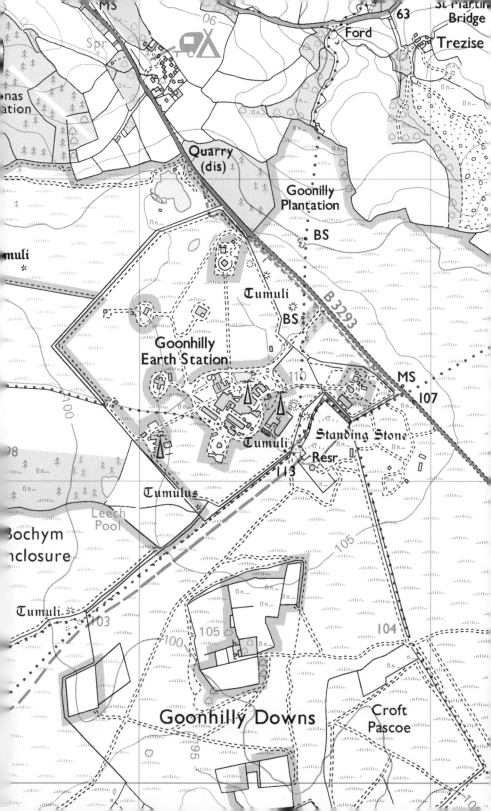

MAP 36

LUMPS AND BUMPS

The map: Landranger at 1:50 000 scale

The story: The start of modern archaeology

Now an Area of Outstanding Natural Beauty, Cranborne Chase survived little changed from its ancient past thanks to its use as a royal hunting ground in Norman times. Neolithic and Bronze Age earthworks litter the landscape around the high chalk ridge along the top of this map. The Romans passed through, too, creating Ackling Dyke. General Pitt Rivers contributed much to modern archaeological methods with his scientific approach to studying the ancient settlements and burial mounds (tumuli) of his estate at Rushmore from the 1880s. He was involved in early efforts to secure the protection of historic sites by the state.

▨ Easy

1. How many of each of the following are marked on the map?
   a. Phone boxes
   b. Places of worship
   c. Milestones ('MS')

2. Which of 'North', 'East', 'South' and 'West' appears furthest north? And which appears furthest south?

▦ Medium

3. What is the difference between the highest and lowest points on the map?

4. How many different words/phrases appear in a Gothic font, indicating a non-Roman historical site? Count plural words/phrases as the same as their singular. Ignore any words that are partially cut off.

▦ Tricky

5. Which locations on the map sound like they might mean the following?
   a. Farm building for a child's game
   b. Feathers from a church
   c. The cow controlled the car
   d. Province framework

6. Which word on the map is also the name of a country?

▦ Challenging

7. Given that the scale of the map is 1:50 000, what is the maximum distance between any two triangulation pillars (blue triangles with a dot inside) as the crow flies, to the nearest half-kilometre?

8. Can you find places on the map whose names are anagrams of the following phrases? Ignore the spaces and punctuation, which may differ from those in the place names.
   a. MOWN TO CUSTOM COD
   b. SO GRAND TOWN
   c. EXPEL DISHY NANNY

MAP
37

BARBARIANS AT THE GATE

The map: Historical Map (1972) enlarged to 1:25 000 scale

The story: Printing Hadrian's Wall

From the earliest days of Ordnance Survey there have been maps highlighting the ancient past. This map of Hadrian's Wall was published in 1972 and was a good example of strip mapping, where the long linear extent of the wall was cut into sections to fit onto the sheet. A 'modern' map base was greyed out to give location context and crude hill shading added to help enhance the landscape over which the wall ran. With a limited palette of spot colours to print with, the archaeological overprint used black for physical remains and red for 'site of'. Text and symbols help finish the picture, but only if you have a legend!

Easy

1. What is the difference in height between the highest and lowest altitudes marked on the map?

2. How many more symbols for springs ('Spr'/'Spring') than wells ('W'/'Ws') are marked on the map?

Medium

3. How many times does 'Warden' appear on the map?

4. The name of which US president appears on the map?

Tricky

5. Can you find two pairs of places on the map that each start with words that mean the opposite of one another?

6. Which locations on the map sound like they could mean the following?
   a. Raw scorch
   b. Brush store
   c. A place where hair-stylers live

Challenging

7. Given that the scale of the map is 1:25 000, how far is the southernmost well from the westernmost well as the crow flies, to the nearest kilometre?

8. Can you find two places on the map which each begin with the name of a mammal? And can you find two places which each begin with the name of a bird?

NETWORK MAPS

Top Trig Trivia

1. Over 6,500 trig pillars were built for the re-triangulation of Great Britain, and around 6,000 are still standing. In total, the re-triangulation had more than 30,000 coordinated points. The modern OS Net network performs the same function with just 110 points.

2. Measuring angles by eye from a trig pillar meant the re-triangulation relied on good weather. No wonder it took until 1962 to complete! Modern GNSS surveying works in all weathers and is available 24 hours a day.

3. Like an iceberg, a large part of the trig pillar is hidden below the surface.

4. The highest trig pillar, unsurprisingly, sits on the top of Ben Nevis. The lowest trig pillar is at Little Ouse, sited at -1m!

5. Trig pillars are mostly made of cast concrete, but a few are built from local stone cemented together.

6. In 2016 Rob Woodall completed a thirteen-year 'trig bagging' mission to visit ('bag') all of Britain's trig pillars.

MAP
38

A WALKER'S FRIEND
ON A FOGGY DAY

The map: MiniScale

The story: Primary Trig Pillars

As part of the switch to a national grid by Ordnance Survey in 1936, a re-triangulation was undertaken. Eleven stations were reused from the work done a century earlier (yellow line on map), followed by a mammoth task to fix and build 6,500 iconic 'trig' pillars often in the least accessible places imaginable. Accurate to 20 metres over the entire length of Great Britain, it required good weather and took until 1962 to complete. In total the re-triangulation had over 30,000 coordinated points, compared to the current satellite-based network OS Net (green squares on map) which performs the same function with just 110 points to an accuracy of 3 millimetres.

QUESTIONS

Easy

1. Which of the four compass points (north, south, east, west) does not appear on the map?

2. How many times does 'shire' appear on the map?

Medium

3. How many place names contain colours?

4. Based on the map, what do the words 'beacon', 'gore' and 'horse' have in common?

Tricky

5. Can you find three above-the-neck parts of the body within the names of places on the map?

6. Which places on the map sound like they would be appropriate for each of these activities?
   a. Sewing
   b. Perusing
   c. Writing
   d. Washing

Challenging

7. Which trig pillar location is both 1° south of Rollright and 1° east of Cleeve Hill?

8. Can you find places on the map whose names are anagrams of the following phrases? Ignore the spaces and punctuation, which may differ from those in the place names.
   a. RUB BAR OWL
   b. SCOLD DILIGENT ANT
   c. AMORAL PIGLET ANNOYS

MAP
39

ALL STRUNG UP

The map: MiniScale at 1:1 000 000 scale

The story: The (other) National Grid

In the 1920s, electricity in Britain was generated locally by numerous private companies and town councils. They often used different systems with different standards and it was expensive, resulting in slow adoption. In 1926 the Electricity Supply Act was passed to connect 122 of the most efficient power stations and establish a national grid. Over 4,000 miles of overhead transmission lines were required, strung up on giant pylons that weren't always welcomed. The grid successfully united the electrical system and, by the Second World War, two-thirds of homes were connected.

▨ Easy

1. How many place names start with a colour? And can you find a place that is also the name of a pattern?

2. How many locations on the map are hyphenated?

▧ Medium

3. Several places on the map start with compass points. Which are the most northerly, easterly, southerly and westerly of each of these?

4. How many times does 'ham' appear within a word on the map?

▨ Tricky

5. How many US state capitals appear on the map?

▨ Challenging

6. Many of the places on the map can be split into two English words. For example, Redditch could be split into red + ditch. Can you find further two-word places, where each word has been replaced by a synonym?
   a. Decompose + she
   b. Shed + fastener
   c. Wheat + divider
   d. Provide + orifice
   e. Bewitch + meat
   f. Sacred + leading
   g. Mammals + inter
   h. Association + crossing
   i. Tease + touching

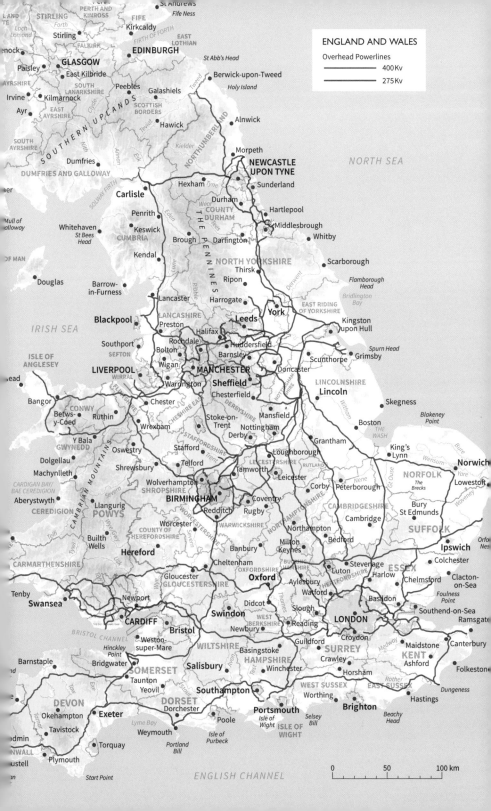

MAP
40

A WALKING BOOT CHALLENGE

The map: MiniScale at 1:1 000 000 scale

The story: National Trails

Of the sixteen National Trails and twenty-two Scottish Great Trails, Offa's Dyke Path offers the closest thing to its designation. Opened in 1971, it closely follows the modern England–Wales border in large parts and the linear earthworks after which it is named. The dyke was constructed along the western edge of the Kingdom of Mercia, with its defensive ditch on the Welsh side. Although largely attributed to King Offa of Mercia, some parts have been dated back to an earlier post-Roman period. Glyndŵr's Way was given National Trail status in 2000 to mark the third millennium and the 600th anniversary of Owain Glyndŵr's rebellion in 1400. The map itself has a story as one of the first maps created using early desktop publishing software in the 1990s, and is still electronically hand-drawn and updated by the cartographer behind this book.

▨ Easy

1. How many aeroplane symbols appear on the map?

2. In how many different colours are numbers printed on the map?

▨ Medium

3. How many times does 'port' appear on the map, including within other words?

4. Which place name on the map contains within its centre a nine-letter brand of car, but with the brand's central letter missing?

▨ Tricky

5. The name of which capital city other than Cardiff appears within the name of a location on the map? And which US state capital?

6. What do the words 'quay', 'castle', 'port' and 'town' have in common, and which is the odd one out?

▨ Challenging

7. What's the height difference between the highest and lowest peaks marked on the map? How far apart are they as the crow flies, to the nearest ten kilometres?

8. Which places on the map could solve the following cryptic crossword clues?
   a. Prelate's fort constructed from chess pieces
   b. Cease to cover short secreting structure in country
   c. Pod members heard land

SOLUTIONS

MAP
I

1. Three

2. Otford

3. Noah's Ark

4. They're all the names of farms on the map

5. Austin

6. Week and Seven Oaks, since a week has seven days

7. a. Warren House; b. Dunstall; c. Steers Hill; d. Tanyard

8. Six (lower, upper, little, east, green, black) – or seven if you include 'Gt' as 'great'

MAP
2

1. 437 (in 'A437')

2. Four

3. bed + font, chat + tern, hat + ton, kempt + on, and sun + bury

4. Five, in the first complete row down, with 'Schs' marked next to two schools in two boxes

5. a. Heathrow; b. Ashford; c. Hanworth; d. Feltham

6. 'Slow': the others all appear at the end of a word that immediately precedes 'Park', but 'slow' appears just before 'Heath', i.e. Hounslow Heath compared to Hanworth Park, Kempton Park and Cranford Park

7. 8km

8. a. North Hyde; b. Queen's Building; c. Ashford Common

MAP
3

1. Brazils

2. Six

3. Honeywood Farm

4. Two

5. Brandy Hole

6. In Cock Clarks, just north of the phone box

7. Four (in Great Whitmans, Great War Aerodrome Museum, Great Canney and Great Hayes)

8. 4km

MAP
4

1. Four

2. a. Crofton Locks; b. Dodsdown; c. Wilton Water

3. 171m

4. They all precede 'on' in names on the map

5. Six: short dashes (footpath), long dashes (bridleway), diamonds (recreational route), circles (other route with public access), half crosses (restricted byway) and crosses (byway open to all traffic)

6. Manor Farm

7. Beam Engines

MAP
5

1. Cowplain

2. North End is south of Southwick

3. Twelve

4. 3,053 (A3056 – A3)

5. They all precede the word 'end' in a place name on the map

6. Five: duck, fish, horse (in the fairground symbol), elephant and deer (head with antlers)

7. Three

8. 3km

MAP
6

1. a. Three; b. Two; c. Seven

2. Eastmoor Street and Westmoor Street

3. Camel Road and Swan Road (and, arguably, Herringham Road)

4. Silvertown

5. Queen Victoria and Albert, in 'Royal Victoria Dock' and 'Royal Albert Dock'

6. All four words are followed by 'way' somewhere on the map. However, 'subway' and 'slipway' are single words, whereas 'Festoon Way' and 'Unity Way' are two separate words

7. 750m

8. 120m

MAP
7

1. North (indicated by 'Hammersmith Bridge')

2. Southwest, in 'London and Southwestern Railway'

3. Cleveland (below and to the left of 939 781)

4. Beverley Brook

5. Hammersmith, Hyde Park Corner and Richmond

6. a. Soap Works; b. Ball Alley; c. Lower Common; d. Mill Lodge

MAP

8

1. Sunday

2. 'Bank', which appears five times (one in 'Banks'), compared to 'back' which appears three times

3. Walker Street

4. 'Queen' and 'Albert'

5. Christ Church

6. a. Slip; b. Vicarage; c. Springfield; d. Metropole

7. a. Ward – 'draw back' = 'draw' written backwards, for 'ward'; 'charge' = straight definition of 'ward' (as in someone to be looked after);

   b. Promenade – 'Walk' = 'promenade'; 'for' = 'pro'; 'males' = 'men'; 'with half decade' = with half the word 'decade', i.e. 'ade'

   c. Tabernacle – 'Confused a celebrant' = an anagram ('confused') of 'a celebrant' to give 'tabernacle'; 'for host storage?' is a literal definition of what a tabernacle is for

   d. Ordinary – 'Regular' = 'ordinary'; 'or' = 'or'; 'variable' = 'y' (as in a maths plot); 'following currency' = written after 'dinar'.

MAP
9

1. ◯ = 1a ⬤ = 1b ◯ = 1c (see map below)

2. 10m, alongside the River Tweed

3. They all precede the word 'Ho' in places on the map

4. West: the other three are in North Sea/Northumberland Coast Path, East Ord and South Ord

5. Seven: eye, head, mouth, shin, shank, thumb and ear – e.g. in Needles Eye, Sharpers' Head, Tweedmouth, Redshin Cove, Letham Shank, Northumberland Road Coast Path and Bear's Head

6. a. Unthank; b. Borewell; c. Ramparts; d. Billylaw

7. 3km

8. a. Berwick-upon-Tweed; b. Heatherytops; c. Devil's Causeway

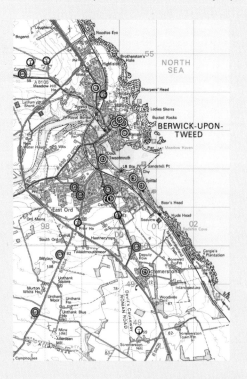

MAP
10

1. Five

2. Fourteen

3. Green and white

4. dun + stable, high + bury, long + spoons, rings + hall, sallow + spring, sew + age and stud + ham

5. 'J'

6. Two

7. 2km

8. a. Homes and Lion; b. Icknield and Tower; c. Bison and Trail; d. Sewage and Tree

MAP
11

1. 345 (in 'A345')

2. Bethnal

3. Two

4. Roundabout

5. 375m

6. Home Farm

7. They're all the middle word of a three-word place name: Huish Down Farm, Huish Hill House and Mid Wilts Way

8. Rainscombe Farm

MAP

12

1. Pennine Way

2. Three (in Upper Red Brook, Nether Red Brook and Red Brook)

3. Jacob's Ladder

4. They all follow the word 'kinder' in place names on the map

5. a. Swine's Back; b. Mermaid's Pool; c. Grouse Butts; d. Wool Packs

6. 1km

MAP

13

1. Four

2. Cattle Market

3. Eight (Mill Lane Farm, Rushmeadow Farm, Dairy Farm, Brook Farm, Manor Farm x2, Fern Farm and Hill Farm)

4. 47 (in 'A47')

5. a. Zero right turns (from the restricted byway on North Runcton Common to the one on Rectory Lane); b. Zero left turns (from the restricted byway on Rectory Lane to the northernmost one, near the Trading Estate)

6. 1,400m

7. Six (in West Winch Common, West Winch, West Winch CP, North Runcton, North Runcton Common and North Runcton CP)

8. a. Chequers Lane; b. Puny Drain; c. The Lodge

MAP
14

1. Four

2. Otterhampton CP

3. Five

4. The Island

5. Cobb's Leaze Clyce/Cobb's Leaze Rhyne

6. a. Collards Farm; b. Stert Flats; c. Pawlett Level

7. 500m

MAP
15

1. Three

2. 'Castle', as in Castle Carrock (the name of a village), Appleby Castle and Brougham Castle

3. a. Raise; b. Knock; c. Bearsbridge; d. Scale Houses; e. Askham

4. They both precede 'endale'

5. Four (Lazonby, Glassonby, Hunsonby and Langwathby)

6. Eden (River Eden)

7. a. Knock – two straight definitions: 'Criticise' = 'knock'; 'rap' = 'knock'

   b. Chimneys – 'Cold' = 'c'; 'guy' = 'him'; 'with new' = 'n'; 'muddled yes' = an anagram ('muddled') of 'yes' to give 'eys'; 'for extraction vents' means that it defines 'extraction vents', to give chimneys

   c. Whale – 'Wide' = 'w'; 'healthy ' = 'hale'; 'large mammal' = 'whale'

MAP
16

1. Ganu Mòr, at 911m (the highest is 927m)

2. Newton

3. Rispond

4. Quinag, at 809m

5. Badcall

6. Cape Wrath

7. 39km

8. a. Achiemore; b. Kylestrome; c. Aultanrynie

MAP
17

1. Four: 'Cairn', 'Settlement', 'Copper' and 'Mine'

2. 1085m and 119m

3. Seven: walks or trails (footprint), parking ('P'), nature reserve (duck), campsite (tepee), picnic site (folding table), visitor centre ('V') and viewpoint (five angled beams)

4. Devil's Kitchen

5. They are all preceded by 'Llyn', the Welsh word for 'lake'

6. a. Roman Camp; b. Summit; c. Miners' Track

7. 1750 metres

8. a. Glyder Fach; b. Yr Ole Wen; c. Elidir Fawr

MAP

18

1. Londonthorpe/Londonthorpe Wood

2. 6,403 (in 'B6403')

3. The one in Grantham, just south of Hall's Hill (the nearest phone box is in Old Somerby)

4. You could play golf or visit a National Trust site, but no castle is marked – a golf course/links is shown just east of The Mill, and a National Trust site is marked in Belton Park

5. The moat by Old Somerby

6. The reservoir south of Alma Wood, marked 'Resr'

7. Manor Farm

8. Belton Park ('belt on park')

MAP
19

1. Four

2. a. Skirnawilse; b. Codlafield; c. Hedlicliv

3. Five (in Hamnafield, Da Tooñ Haṁ, Ham Little, Hamar and Da Hametoon)

4. Da Sheepie and Da Baas o Stremness or Da Head o da Baa ('sheep' and 'baa')

5. Harrier

6. a. Summons Head; b. Braidfit

7. Rossie's Loch

8. 51

MAP
20

1. Five

2. Four (in Walcott Green, Whimpwell Green, Green Farm and Green Lane)

3. Three (Willow Farm, Oak and Holly Farm)

4. Short Lane

5. They are all names of farms

6. No right turns at junctions are required

7. Blacksmiths House

8. Coronation Close

MAP
21

1. Four (in Greengates Quarry and Old Quarry x3)

2. Three

3. Fifteen (Black, Crushing, Engine, Gill, Greengates, House, Limekiln, Mill, Old, Quarry, Saddlebow, Sheepfold, Tramway, Trough and Tunnel)

4. 1,148

5. With a letter added to the front of them, they are all words that appear on the map (Gill/Mill, Black, Trough and Crushing)

6. 'Green' in Greengates and 'Lime' in Limekiln

7. 'W'

MAP

22

1. Three

2. 494

3. 'Ford', and it appears twice

4. Nine (in Gowanburn, Dry Burn, Bellingburn Head, Plashetts Burn, Belling Burn, Belllingburn, Lewisburn, Whickhope Burn and Little Cranecleugh Burn)

5. Mounces, Otterstone Lee and Millstone Crag

6. Hawkhirst, Otterstone, Bullcrag and Cranecleugh

7. Stower Hill

8. The Law

MAP
23

1. Six

2. 557 (in 'B557')

3. Milton Keynes

4. Woughton on the Green

5. The three places are Simpson (as in *The Simpsons*), Walton (as in *The Waltons*) and Springfield (location of *The Simpsons*)

6. a. Cross End; b. Little Brickhill; c. Tickford Park; d. Tongwell

7. Peartree Br. (as in 'And a partridge in a pear tree' from 'The Twelve Days of Christmas')

8. Old Covert and New Covert

MAP
24

1. Mars (in West Thurrock Marshes)

2. 'Low' in Low Well Wood and 'High' in Mean High Water

3. 'Z' is the only letter that doesn't appear on the map; the first letter not to appear capitalised is 'U'

4. 'V'

5. Seven: pink, white (or transparent), green, orange, black, red and blue

6. brick + barn, broom + hill, cause + way, lake + side, and me + an

7. 3,750 m

8. a. Karting – 'Racing' = straight clue for 'Karting; 'leader' = 'king'; bottling (surrounding) 'drawing perhaps' = 'art', i.e. kart**ing**

   b. Caravan – 'Two vehicles' = 'car' and 'van'; 'join around a' = surround the letter 'a' giving '**car**a**van**'; 'group of people' = straight clue for 'caravan', with the question mark implying 'perhaps'

   c. Complex – 'Real and imaginary' = straight clue for 'complex', as in the definition of a complex number; 'group of building' is a second straight clue for 'complex'.

MAP
25

1. Five

2. Nine

3. 'Water' and 'Fire' (in Water Tower and Fire Hill)

4. Red House Farm and Blackman's Head, and Dovercourt and Crane's Hill

5. Orwell

6. They are all followed by 'Ho' in places on the map

7. 'N'

8. a. Vicarage; b. Harwich Harbour; c. Trimley St. Mary

MAP
26

1. Abbey Road

2. Brookvale Park

3. Eight

4. 'Park' (as in Park Place, Park Road and Park Approach)

5. Waterworks Street shares its name with 'Water Works' on a Monopoly board

6. a. River Tame; b. Cuckoo Bridge

7. YMCA

8. a. Gladstone Road; b. Bourne Road

MAP

27

1. Six

2. Three stadiums and three arenas

3. River Lea or Lee

4. They all follow the word 'Olympic' in the name of a location

5. a. Biggerstaff Road; b. High Meads Junction; c. Temple Mills; d. Iceland R; e. Channelsea River; f. Major Road

6. 275m

7. 'Velodrome' contains 'Rome'

8. a. Stratford; b. Channel Tunnel Rail Link; c. Northern Outfall Sewer

MAP
28

1. 185m

2. 'Mells' – appears six or seven times (seven if you include Mellsgreen), while 'Vobster' only appears five times

3. They are all the names of farms

4. They all follow the word 'the' on the map

5. 'Up' e.g. in Upton's Piece, and 'Down' e.g. in Shears Down Farm

6. a. Coldbath Plantation; b. Finger Farm; c. Shears Down Farm

7. 5.5km

8. a. Fish Pond – 'Keen on' = 'fond'; 'submerging' (surrounding) 'broken ship' = an anagram (broken) of 'ship' to give **F**ish p**ond**; 'for swimmers?' is a whimsical straight definition of a fish pond

   b. Foxholes – 'Throw' = definition of 'fox', as in to confuse; 'jams' = a definition of 'holes', as in bad situations; 'in' is a linking word to the definition; 'dens' = straigh definition of foxholes.

   c. Dismantled Railway – 'Woman's' is 'Di's' (i.e. a woman called Di); 'guy' = 'man'; 'with' is a joining word to say to add on to what's next; 'a weird tally confused' = an anagram ('confused') of 'a weird tally', to give 'tled railway'; 'previous tracks?' = straight clue for 'dismantled railway'.

MAP
29

1. North and east (in North Wilts Branch, Eastcott Lodge and East Street)

2. 345.1

3. They are all followed by 'Street' on the map

4. a. Three; b. Eight; c. Six; d. Seven

5. Six: Bristol Street, London Street, Bath Street, Exeter Street, Reading Street and Chester Street

6. Henry Street and Catherine Street

7. 800m

8. a. Drill Hall – 'Practice' = 'drill'; 'passage' = 'hall'; 'infantry tested here?' is a straight definition of 'drill hall'

   b. Lock House – 'insect' = 'louse'; 'climbing over broken hock' = written across 'climbing over' an anagram ('broken') of hock, to give **lo**ck ho**use**; 'for flow controller?' is a quirky but striaght definition of what a lock house is – a place for the flow controller

   c. Engine Sheds – 'Motor' = 'engine'; 'moults' = 'sheds'; 'for trains?' defines 'engine sheds'

   d. Brick Works – 'Brother' = 'br'; 'kicks' around = an anagram of 'kicks' (around); 'over' means 'written across'; 'back line' = 'row' written backwards, so giving **Br**ick **Wor**ks; 'block factory' is a straight definition of 'brick works'

MAP
30

1. Seven

2. Two: Penallta Country Park and Business Park

3. 47 (the traffic-free national cycle network route number shown in the red box), and 4254 (in 'B4254')

4. Six: short dashes (footpath), long dashes (bridleway), diamonds (recreational route), green circles (other route with public access), crosses (byway open to all traffic) and orange circles (traffic-free cycle route)

5. a. Green (in Green Acres), or red (in Tredomen); b. Green (in Green Acres), or red (in Tredomen); c. Oman, in 'Roman Forts'

6. Nine: Penpedairheol, Penybryn, Penallta, Penalltau Road, Penallta Country Park, Penalltau-fawr, Penalltau-isaf, Penywaun and Pen-yr-heol-fawr

7. 4km – the furthest north place of worship is to the east of the Post Office at the top right, and the furthest south place of worship is below the college at the bottom right

8. a. Spout; b. Trial; c. Shaft

MAP
31

1. a. Six; b. Seven; c. Seven

2. 270m (325–55)

3. Five

4. Six: well ('W'), spring ('Sprs'), reservoir, spout, brook and stream

5. Argoed Farm (which contains 'Argo')

6. Nine: horse riding (horseshoe), public house (beer tankard), boat trips (boat on water), public toilets (man and woman), parking ('P'), World Heritage Site (square in a circle), visitor centre ('V'), picnic site (folding table) and country park (stylised family)

7. 3,500m

8. 'W'

MAP 32

1. Three (Pule Hill, Beacon Hill and Shaw Hill)

2. Two

3. Dunkirk

4. Six (in the square that is sixth down and third across, counting partial squares)

5. The words can be paired by joining them with 'ing' to form the names of two towns on the map: Beggarington and Illingworth; this makes the pairs 'beggar' and 'ton', and 'ill' and 'worth'

6. a. Ovenden; b. Bank Top; c. Queensbury; d. Holdsworth

7. 300m

8. e. Ambler Thorn; f. Catharine Slack; g. Stump Cross

MAP

33

1. Five

2. Six (in Meadows Way, Meadow Lane x2, Meadow La[ne], Meadow C[lose] and Meadow Gr[ove])

3. Three: football, cricket and tennis (in 'Football Ground', 'Trent Bridge Cricket Ground'/ 'Cricketeers Ct' and 'Little Tennis Street')

4. Five (William Road, George Road, Henry Road, Edward Road and Victoria Road)

5. a. Refuse Incinerator; b. Melton Road; c. Millicent Road ('road' to 'mill' 'I' 'cent')

6. London Road (capital of the UK) and Victoria Road (capital of the Seychelles)

7. 1,600m

8. d. Trent Bridge Cricket Ground; e. Environment Agency; f. Sneinton Junction

MAP

34

1. Five

2. Three: a black cross (a place of worship), a black rectangle with a cross (a place of worship with tower) and a black circle with a cross (a place of worship with spire, minaret or dome)

3. They all start the name of a farm (Cliff Farm, Old Home Farm, Home Farm, Gainsthorpe Farm, Quarry Fields Farm, Northwood/Northcliffe Farm and Springcliff Farm)

4. a. Five (Cliff Farm, North Cliff Road, Northcliffe Farm, Kirton Cliff and Springcliff Farm); b. Five (Kirton Tunnel, Kirton Lindsey Station, Kirton in Lindsey, Kirton Cliff and Kirton in Lindsey CP); c. Three (Gainsthorpe Road, Gainsthorpe Medieval Village and Gainsthorpe Farm)

5. Ermine (a stoat) in 'Ermine Street', and fox in 'Fox House'

6. 1,385, made up of 1,400 (in 'B1400') – 15 (in 'A15')

7. 3km

8. d. Works – 'Functions' = 'works'; 'right' = 'r'; 'within' = written inside; 'pans' 'works' to give works

   e. Garden Centre – 'Recent danger damaged = anagram ('damaged') of 'recent danger' to give 'garden centre'; 'plant store' = definition of 'garden centre'

MAP 35

1. Five

2. 113

3. Three: spring ('Spr'), reservoir ('Resr') and pool ('Pool')

4. Bochym Enclosure

5. Six (in 'Bridge', 'Dobnas', 'BS' x2, 'B3293' and 'Bochym')

6. Nine (in 'Quarry', 'Leech', 'Pool', two in 'Goonilly', and two in 'Goonhilly' x2)

7. 400m

8. a. Croft + Station; b. Ford + Enclosure; c. Downs + Standing

MAP
36

1. a. Five; b. Six (five with towers and one without); c. Five

2. 'East' appears furthest north in East Combe Wood, and 'North' appears furthest south in North Fm.

3. 189m (248–59)

4. Ten: Field System, Tumulus/Tumuli, Winkelbury, Fort, Settlement, Scrubbity Barrows, Enclosure, King John's Ho, Long Barrow(s) and Cursus

5. a. Chase Barn; b. Chapel Down; c. Ox Drove; d. Shire Rack

6. Canada (in 'Canada Fm')

7. 8.5km

8. a. Woodcutts Common; b. Garston Down; c. Sixpenny Handley

MAP

37

1. 525m (700–175)

2. Five more springs

3. Four (in Warden, High Warden, Nether Warden and Warden Hill)

4. Lincoln (in Lincoln Hill)

5. High Barns and Low Barns, and High Brunton and Low Brunton (also, High Warden and Nether Warden are partial opposites)

6. a. Red Burn; b. Broom Park; c. Acomb House

7. 4km

8. Cowper Hill and Horseley Wood, and Ravenside and Cocklaw Quarry

MAP
38

1. East (north in Northampton, south in Southampton/South Gloucestershire, and West in West Berkshire/Westbury Down)

2. Eleven: Worcestershire, Warwickshire, Northamptonshire, County of Herefordshire (only partially visible), Gloucestershire, Buckinghamshire, Oxfordshire, South Gloucestershire, West Berkshire, Wiltshire and Hampshire

3. Four: Redditch, White Horse Hill, Wingreen and Blackdown Pillar

4. They all precede 'Hill'

5. Nose in Dunnose, mouth in e.g. Bournemouth, and head in e.g. St Alban's Head

6. a. The Needles; b. Reading; c. Inkpen; d. Bath

7. Butser, at 51°N and 1°W

8. a. Bulbarrow; b. Liddington Castle; c. Royal Leamington Spa

MAP
39

1. Four (Greenock, Blackpool, Whitehaven and Redditch); Paisley is a pattern

2. Seven: Berwick-upon-Tweed, Barrow-in-Furness, Betws-y-Coed, Stoke-on-Trent, Clacton-on-Sea, Southend-on-Sea and Weston-super-Mare

3. The furthest north is East Lothian; the furthest east is Southend-on-Sea; the furthest south is Southampton; the furthest west is North Ayrshire

4. Nineteen: Witham, Northamptonshire, Thames, Wrexham, Southampton, Hexham, Nottingham, Wolverhampton, Birmingham, Northampton, Cheltenham, Okehampton, Horsham, Grantham, Durham, County Durham, Hampshire, Buckinghamshire and Nottinghamshire

5. Two: Lincoln (Nebraska) and Boston (Massachusetts)

6. a. Rother; b. Barnstaple; c. Cornwall; d. Plymouth; e. Hexham; f. Holyhead; g. Shrewsbury; h. Guildford; i. Taunton

MAP
40

1. Three

2. Seven: white (e.g. M53), yellow (e.g. A55), pink (e.g. A489), brown (e.g. 1085m), green (e.g. A483), blue (M49) and black (motorway junctions, and in the scale)

3. Eleven: Ellesmere Port, Porthmadog, Portishead, Burry Port, Port-Eynon, Porthcawl, Aberporth, Port Talbot, Neath Port Talbot and Newport x2

4. Llandovery (contains 'Land Rover' without the central 'r')

5. Amman (in Ammanford/Brynamman, capital of Jordan) and Montgomery (capital of Alabama)

6. All of the words come after 'New' in a location on the map, but 'quay' is the only one that's a separate word (New Quay, Newcastle Emlyn, Newport and Newtown)

7. 199m height difference and 140km apart (the peaks are Snowdon and Pen y Fan)

8. a. Bishops Castle – 'Prelate's fort' = straight definition of 'bishop's castle'; 'constructed from chess pieces' = bishops + castle, i.e. the word is made up of chess pieces

   b. England – 'Cease' = 'end'; 'to cover short secreting structure' = written across ('to cover') 'gland' (a 'secreting structure'), shortened ('short') by one letter; 'in country' meaning that the answer can be found in a word defined by 'country'

   c. Wales – 'Pod members' is 'whales', and 'heard' means that it sounds like that, i.e. 'Wales'; 'land' = straight definition for 'Wales'

ACKNOWLEDGEMENTS

Thank you to the many people who have worked hard
to make this book happen, including:

Nick Giles, Keegan Wilson, Gemma Bell, Ed Graham Campbell, Jo Lines,
Jim Goldsmith, Nic Hamilton, Carolyne Lawton and Maria Court. Thanks to
Zoe Walker for her meticulous checking.

Special thanks to Mark Wolstenholme, who put a lot of hours into this project,
and Paul McGonigal, for steering the book through at
Ordnance Survey.

ORDNANCE SURVEY MAPS

You've (hopefully…) completed the puzzles in this book. Congratulations – why not celebrate with a cup of your favourite hot beverage, treat yourself to a biscuit and then celebrate further by getting outside. Our index tells you which of our iconic paper maps you will need to visit the locations featured in this book. Some sites are admittedly more interesting than others (no offence, Spaghetti Junction), but taken as a whole they are a reminder of what an interesting and varied country we live in.

And if you haven't yet completed all the puzzles then by all means take this book with you on your next adventure. Clear the head, stretch those legs, get some fresh air in the lungs. Perhaps the Great British Outdoors will be your inspiration!

OS Explorer

1: 25 000 (4cm to 1km or 2½ inches to 1 mile)

For outdoor enthusiasts who want to make the most of all Great Britain has to offer, OS Explorer is the nation's most popular leisure map. It features footpaths, rights of way and open access land, and is our recommended map for walking, running and horse riding. The OS Explorer map covers a smaller area than the OS Landranger map, but presents the landscape in more detail, aiding navigation and making it the perfect accompaniment on an adventure. It also highlights tourist information and points of interest, including viewpoints and pubs. Use it to plan your route but don't leave it at home.

OS Landranger

1: 50 000 (2cm to 1km or 1¼ inches to 1 mile)

For dedicated or part-time explorers who want to maximise their often limited free time, OS Landranger is a map that aids the planning of the perfect short break in Great Britain, and is a vital resource for identifying opportunities for experiences in both towns and countryside. OS Landranger displays larger areas of the country than OS Explorer, making it more suitable for touring an area by car or by bicycle, helping you access the best an area has to offer, and all its variety.

OS SHEET INDEX

	Puzzle map	Feature ref	OS map sheet	
1	Kent	TQ 528 592	147	188
2	Feltham, London	TQ 105 735	160, 161	176
3	Stow Maries, Essex	TL 819 002	175, 183	168
4	Crofton, Wiltshire	SU 257 620	157	174
5	Spithead, Hampshire	SZ 650 940	OL3, OL29	196
6	Thames Barrier	TQ 414 795	162	177
7	Barnes, London	TQ 224 771	161	176
8	Blackpool	SD 306 365	286	102
9	Berwick-upon-Tweed	NU 000 530	346	75
10	Whipsnade, Central Bedfordshire	SP 996 176	181	166
11	Oare, Wiltshire	SU 154 636	157	173
12	Kinder Scout, Peak District	SK 085 875	OL1	110
13	West Winch, Norfolk	TF 634 155	236	132
14	River Parrett, Somerset	ST 267 446	140	182
15	Cross Fell, Cumbria	NY 687 343	OL31	91
16	Cape Wrath, Highland	NC 224 656	446	9
17	Pass of Llanberis, Gwynedd	SH 627 567	OL17	115
18	Cold Harbour, Lincolnshire	SK 952 347	247	130
19	Foula, Shetland Islands	HT 951 375	467	4
20	Happisburgh, Norfolk	TG 380 310	252	133
21	Greengates, County Durham	NY 933 235	OL31	92
22	Kielder Forest, Northumberland	NY 663 901	OL42	80
23	Milton Keynes	SP 888 393	192	152
24	Lakeside, West Thurrock	TQ 583 789	162	177
25	Felixstowe, Suffolk	TM 282 336	197	169
26	Gravelly Hill, Birmingham	SP 092 903	220	139
27	Stratford, London	TQ 376 840	162	177
28	Vobster, Somerset	ST 709 493	142	183
29	Swindon	SU 146 850	169	173
30	Penallta, Monmouthshire	ST 131 954	166	171
31	Pontcysyllte Aqueduct, Wrexham	SJ 270 416	256	117
32	Halifax	SE 090 285	288	104
33	Nottingham	SK 585 382	260	129
34	Gainsthorpe Medieval Village	SE 954 011	281	112
35	Goonhilly Earth Station	SW 723 212	103	204
36	Cranborne Chase, Dorset	ST 972 179	118	184
37	Hadrian's Wall	NY 907 704	OL43	87

38 to 40 – network maps not listed

Share your adventures and
puzzle-solving with us:

os.uk/blog
@ordnancesurvey
@OSLeisure
@osmapping

Just tag #OSpuzzlebook